Stark Broadening of Spectral Lines in Plasmas

Stark Broadening of Spectral Lines in Plasmas

Special Issue Editor

Eugene Oks

MDPI • Basel • Beijing • Wuhan • Barcelona • Belgrade

MDPI

Special Issue Editor
Eugene Oks
Auburn University
USA

Editorial Office
MDPI
St. Alban-Anlage 66
4052 Basel, Switzerland

This is a reprint of articles from the Special Issue published online in the open access journal *Atoms* (ISSN 2218-2004) in 2018 (available at: https://www.mdpi.com/journal/atoms/special_issues/stark_broadening_plasmas)

For citation purposes, cite each article independently as indicated on the article page online and as indicated below:

LastName, A.A.; LastName, B.B.; LastName, C.C. Article Title. *Journal Name* **Year**, *Article Number, Page Range.*

ISBN 978-3-03897-455-0 (Pbk)
ISBN 978-3-03897-456-7 (PDF)

Cover image courtesy of Eugene Oks.

Contents

About the Special Issue Editor

Eugene Oks received his Ph.D. degree from the Moscow Institute of Physics and Technology and, later, the highest degree of Doctor of Sciences from the Institute of General Physics of the Academy of Sciences of the USSR by the decision of the Scientific Council led by the Nobel Prize winner, academician A.M. Prokhorov. Oks worked in Moscow (USSR) as the head of a research unit at the Center for Studying Surfaces and Vacuum, then at the Ruhr University in Bochum (Germany) as an invited professor, and, for the last 28 years at the Physics Department of Auburn University (USA) in the position of Professor. He has conducted research in five areas: atomic and molecular physics, plasma physics, laser physics, nonlinear dynamics, and astrophysics. He founded/co-founded and developed new research fields, such as intra-Stark spectroscopy (new class of nonlinear optical phenomena in plasmas), masing without inversion (advanced schemes for generating/amplifying coherent microwave radiation), and quantum chaos (nonlinear dynamics in the microscopic world). He also developed a large number of advanced spectroscopic methods for diagnosing various laboratory and astrophysical plasmas, methods that were then used and are used by many experimental groups around the world. He published over 400 papers and 5 books, the latest one, published in 2017, being titled "Diagnostics of Laboratory and Astrophysical Plasmas Using Spectral Lineshapes of One-, Two-, and Three-Electron Systems". He is the Chief Editor of the journal *International Review of Atomic and Molecular Physics*. He is a member of the Editorial Boards of the two other journals: *International Journal of Spectroscopy* and *Open Journal of Microphysics*. He is also a member of the International Program Committees of the two conferences: "Spectral Line Shapes" and Zvenigorod Conference on "Plasma Physics and Controlled Fusion".

Preface to "Stark Broadening of Spectral Lines in Plasmas"

Stark broadening of spectral lines remains an important tool for spectroscopic diagnostics of various types of laboratory and astrophysical plasmas. This is because laboratory and astrophysical plasmas contain various types of electric fields, such as the ion microfield, the electron microfield, fields of different kinds of electrostatic plasma turbulence, and laser/maser fields penetrating into plasmas. All these kinds of electric fields, differing by their statistical properties, strength, frequency, and possible polarization, cause a variety of types of Stark broadening of spectral lines in plasmas.

Therefore, this research area is very important both fundamentally and practically, the latter being due to the numerous applications of plasmas. The practical applications range from the controlled thermonuclear fusion to plasma-based lasers and plasma sources of incoherent x-ray radiation as well as technological microwave discharges. Spectroscopic diagnostics based on Stark broadening are also indispensable tools for analyzing various astrophysical objects, such as solar plasmas (both for the quiet Sun and in solar flares), flare stars, white dwarfs, and so on.

This book presents the latest research achievements in the area of Stark broadening of spectral lines in plasmas. It opens with three reviews describing the latest advances in analytical theory (with applications to various laboratory and astrophysical plasmas), in experimental studies of relativistic laser–plasma interactions, and in laboratory experiments facilitating the analysis of white dwarfs in astrophysics.

These reviews are followed by seven original research papers devoted to a number of issues concerning Stark broadening of spectral lines in plasmas. They cover various experimental studies of laser-produced plasmas and theoretical studies presenting, in particular (but not exclusively), a new method for measuring ultra-strong magnetic fields, an improved method for measuring the electron density, as well as an advanced description of the radiative transfer.

Eugene Oks
Special Issue Editor

atoms

MDPI

Review

Review of Recent Advances in the Analytical Theory of Stark Broadening of Hydrogenic Spectral Lines in Plasmas: Applications to Laboratory Discharges and Astrophysical Objects

Eugene Oks

Physics Department, 206 Allison Lab, Auburn University, Auburn, AL 36849, USA; goks@physics.auburn.edu

Received: 26 June 2018; Accepted: 1 August 2018; Published: 3 September 2018

Abstract: There is presented an overview of the latest advances in the analytical theory of Stark broadening of hydrogenic spectral lines in various types of laboratory and astrophysical plasmas. They include: (1) advanced analytical treatment of the Stark broadening of hydrogenic spectral lines by plasma electrons; (2) center-of-mass effects for hydrogen atoms in a nonuniform electric field: applications to magnetic fusion, radiofrequency discharges, and flare stars; (3) penetrating-ions-caused shift of hydrogenic spectral lines in plasmas; (4) improvement of the method for measuring the electron density based on the asymmetry of hydrogenic spectral lines in dense plasmas; (5) Lorentz–Doppler broadening of hydrogen/deuterium spectral lines: analytical solution for any angle of observation and any magnetic field strength, and its applications to magnetic fusion and solar physics; (6) Revision of the Inglis-Teller diagnostic method; (7) Stark broadening of hydrogen/deuterium spectral lines by a relativistic electron beam: analytical results and applications to magnetic fusion; (8) Influence of magnetic-field-caused modifications of the trajectories of plasma electrons on shifts and relative intensities of Zeeman components of hydrogen/deuterium spectral lines: applications to magnetic fusion and white dwarfs; (9) Influence of magnetic-field-caused modifications of trajectories of plasma electrons on the width of hydrogen/deuterium spectral lines: applications to white dwarfs; (10) Stark broadening of hydrogen lines in plasmas of electron densities up to or more than $N_e \sim 10^{20}$ cm^{-3}; and, (11) The shape of spectral lines of two-electron Rydberg atoms/ions: a peculiar Stark broadening.

Keywords: Stark broadening; analytical theory; shapes and shifts of spectral lines; laboratory plasmas; astrophysical plasmas

---

Table of Contents

1. Introduction

Stark broadening of hydrogenic spectral lines remains as an important tool for spectroscopic diagnostics of various types of laboratory and astrophysical plasmas. This is because laboratory and astrophysical plasmas contain various types of electric fields, such as the ion microfield, the electron microfield, fields of different kinds of the electrostatic plasma turbulence, and laser/maser fields penetrating into plasmas. All of these kinds of electric fields, differing by their statistical properties, strength, frequency, and possible polarization, cause a garden variety of the types of Stark broadening of spectral lines in plasmas.

Therefore, this research area is very important both fundamentally and practically, the latter being due to numerous applications of plasmas. The practical applications range from the controlled thermonuclear fusion to plasma-based lasers and plasma sources of incoherent X-ray radiation, as well as technological microwave discharges.

Studies that are related to this research area over its lifetime are too numerous to be listed here in their entirety. So, we refer here to seven books [1–7], published over the last 25 years, and to the references therein.

Our review focuses at latest advances in the *analytical* theory of the Stark broadening of *hydrogenic* spectral lines in plasmas. Studying the Stark broadening of hydrogenic lines is important from the theoretical point of view for two reasons: (1) it is a deeply fundamental problem of the simplest, two-particle bound Coulomb system immersed in a multi-particle Coulomb system of free charges (plasma) exhibiting long-range interactions—as was noted by Lisitsa [8]; (2) hydrogenic atoms/ions possess a higher algebraic symmetry than its geometrical symmetry, thus allowing for significant analytical advances in the Stark broadening problem, and therefore yielding a deep physical insight.

The Stark broadening of hydrogenic spectral lines in plasmas has also a great practical importance. This is because hydrogenic spectral lines are employed for diagnosing plasmas in magnetic fusion and laser fusion machines, as well as in powerful Z-pinches (used for producing X-ray and neutron radiation, ultra-high pulsed magnetic fields), in X-ray lasers, in low-temperature technological discharges for plasma processing, and in astrophysics (especially in solar physics and in physics of flare stars and white dwarfs).

As for focusing at the advances in the analytical theory (versus simulations), the following should be noted. Of course, simulations are important as the third powerful research methodology—in addition to theories and experiments: large-scale codes have been created to simulate lots of complicated phenomena. However, first, not all large-scale codes are properly verified and validated, as illustrated by some well-known failures of large-scale codes (see, e.g., [9,10]). Second, fully-numerical simulations are generally not well-suited for capturing the so-called emergent principles and phenomena, such as, e.g., conservation laws and preservation of symmetries, as explained in [9]. Third, as any fully-numerical method, they lack the physical insight.

2. Advanced Analytical Treatment of the Stark Broadening of Hydrogenic Spectral Lines by Plasma Electrons

The theory of the Stark broadening of hydrogenlike spectral lines by plasma electrons, developed by Griem and Shen [11] and are later presented also in books [12,13], is usually referred to as the Conventional Theory, hereafter CT, also known as the standard theory. (Further advances in the theory of the Stark broadening of hydrogenlike spectral lines by plasma electrons can be found, e.g., in books [5,7] and references therein.) In the CT, the perturbing electrons are considered moving along hyperbolic trajectories in the Coulomb field of the effective charge $Z - 1$ (in atomic units), where Z is the nuclear charge of the radiating ion. In other words, in the CT, there was made a simplifying assumption that the motion of the perturbing electron can be described in frames of a two-body problem, one particle being the perturbing electron and the other "particle" being the charge $Z - 1$.

However, in reality, one have to deal with a three-body problem: the perturbing electron, the nucleus, and the bound electron. Therefore, trajectories of the perturbing electrons should be more complicated.

In paper [14], the authors took this into account by using the standard analytical method of separating rapid and slow subsystems—see, e.g., book [15]. The characteristic frequency of the motion of the bound electron around the nucleus is much higher than the characteristic frequency of the motion of the perturbing electron around the radiating ion. Therefore, the former represents the rapid subsystem and the latter represents the slow subsystem. This approximate analytical method allows for a sufficiently accurate treatment in situations where the perturbation theory fails—see, e.g., book [15].

By applying this method, the authors obtained in [14] more accurate analytical results for the electron broadening operator than in the CT. They showed by examples of the electron broadening of the Lyman lines of He II that the allowance for this effect increases with the electron density N_e, becomes significant already at $N_e \sim 10^{17}$ cm^{-3} and very significant at higher densities. Here, are some details.

The first step in the method of separating rapid and slow subsystems is to "freeze" the slow subsystem (perturbing electron) and to find the analytical solution for the energy of the rapid subsystem (the radiating ion) that would depend on the frozen coordinates of the slow subsystem (in the case studied in [14] it was the dependence on the distance R of the perturbing electron from the radiating ion). To the first non-vanishing order of the R-dependence, the corresponding energy in the parabolic quantization was given by

$$E_{nq}(R) = -\frac{Z^2}{n^2} + \frac{3\,n\,q}{2\,Z\,R^2} \tag{2.1}$$

where n and $q = n_1 - n_2$ are the principal and electric quantum numbers, respectively, of the Stark state of the radiating ion; n_1 and n_2 are the parabolic quantum numbers of that state.

The next step in this method is to consider the motion of the slow subsystem (perturbing electron) in the "effective potential" $V_{eff}(R)$, consisting of the actual potential plus $E_{nq}(R)$. Since the constant term in Equation (2.1) does not affect the motion, the effective potential for the motion of the perturbing electron could be represented in the form

$$V_{eff}(R) = -\frac{\alpha}{R} + \frac{\beta}{R^2}, \quad \alpha = Z - 1 \tag{2.2}$$

For the spectral lines of the Lyman series, since the lower (ground) state b of the radiating ion remains unperturbed (up to/including the order $\sim 1/R^2$), the coefficient β is

$$\beta = \frac{3\,n_a q_a}{2\,Z} \tag{2.3}$$

For other hydrogenic spectral lines, for taking into account both the upper and the lower states of the radiating ion, the coefficient β can be expressed as

$$\beta = \frac{3\,(n_a q_a - n_b q_b)}{2\,Z} \tag{2.4}$$

The motion in the potential from Equation (2.2) allows an exact analytical solution. In particular, the relation between the scattering angle and the impact parameter becomes (see, e.g., book [16])

$$\Theta = \pi - \frac{2}{\sqrt{1 + \frac{2\,m\,\beta}{M^2}}} \arctan \sqrt{\frac{4\,E}{\alpha^2}\left(\beta + \frac{M^2}{2m}\right)} \tag{2.5}$$

Here, E and M are the energy and the angular momentum of the perturbing electron, respectively. Since the angular momentum can be written in terms of the impact parameter ρ as

$$M = m\,v\,\rho \tag{2.6}$$

then a slight rearrangement of Equation (2.5) yielded

$$\tan\left(\frac{\pi - \Theta}{2}\sqrt{1 + \frac{2\,\beta}{m\,v^2\rho^2}}\right) = \frac{v}{\alpha}\sqrt{m^2 v^2 \rho^2 + 2\,m\,\beta} \tag{2.7}$$

In the CT, after calculating the S matrices for weak collisions, the electron broadening operator becomes (*in atomic units*)

$$\Phi_{ab}^{weak} \equiv C \int_{\rho_{we}}^{\rho_{max}} d\rho\,\rho \sin^2\frac{\Theta(\rho)}{2} = \frac{C}{2}\int_{\Theta_{min}}^{\Theta_{max}} d\Theta \frac{d\rho^2}{d\Theta} \sin^2\frac{\Theta}{2} \tag{2.8}$$

where Θ is the scattering angle for the collision between the perturbing electron and the radiating ion and the operator C is

$$C = -\frac{4\,\pi}{3}N_e\left[\int_0^\infty dv\,v^3 f(v)\right] \frac{m^2}{(Z-1)^2}(r_a - r_b{}^*)^2 \tag{2.9}$$

Here, $f(v)$ is the velocity distribution of the perturbing electrons, m is the reduced mass of the system "perturbing electron—radiating ion"; r is the radius-vector operator of the bound electron (which scales with Z as $1/Z$).

So, after solving Equation (2.7) for ρ and substituting the outcome in Equation (2.8), a more accurate expression for the electron broadening operator can be obtained. In [14], to get the message across in the simplest form, the authors solved Equation (2.7) by expanding it in powers of β.

After combining the contributions from weak and strong collisions, the authors obtained the following final results for the electron broadening operator:

$$\Phi_{ab}(\beta) = -\frac{4}{3}\pi N_e(r_a - r_b{}^*)^2\left[\int_0^\infty dv\,\frac{f(v)}{v}\right]\left\{\frac{1}{2}\left[1 - \frac{3}{2}\frac{Z^2(Z-1)^2}{\left(n_a^2 - n_b^2\right)^2 m^2\,v^2}\right] + \log\left[\sqrt{\frac{3}{2}\frac{Z\,v\,\rho_D}{(n_a^2 - n_b^2)}}\sqrt{1 + \left(\frac{Z-1}{m\,v^2\rho_D}\right)^2}\right] + \frac{mv^2\beta}{(Z-1)^2}\left(\frac{\pi^2}{4} - 1\right)\right\} \tag{2.10}$$

for the non-Lyman lines and

$$\Phi_{ab}(\beta) = -\frac{4}{3}\pi N_e r_a^2\left[\int_0^\infty dv\,\frac{f(v)}{v}\right]\left\{\frac{1}{2}\left[1 - \frac{3}{2}\frac{Z^2(Z-1)^2}{n_a^4 m^2\,v^2}\right] + \log\left[\sqrt{\frac{3}{2}\frac{Z\,v\,\rho_D}{n_a^2}}\sqrt{1 + \left(\frac{Z-1}{m\,v^2\rho_D}\right)^2}\right] + \frac{mv^2\beta}{(Z-1)^2}\left(\frac{\pi^2}{4} - 1\right)\right\} \tag{2.11}$$

for the Lyman lines. Here, and below log[...], stands for the natural logarithm.

For determining the significance of the effect of non-hyperbolic trajectories manifested by the third term in braces in Equation (2.10) or Equation (2.11), the authors evaluated the ratio of that term to the first two terms in the same braces

$$ratio = \frac{\frac{3}{2} \frac{mv^2(n_aq_a - n_bq_b)}{(Z-1)^2}\left(\frac{\pi^2}{4} - 1\right)}{\frac{1}{2}\left[1 - \frac{3}{2}\frac{Z^2(Z-1)^2}{(n_a^2-n_b^2)^2 m^2 v^2}\right] + \log\left[\sqrt{\frac{3}{2}\frac{Z\,v\,\rho_D}{(n_a^2-n_b^2)}}\sqrt{1 + \left(\frac{Z-1}{m\,v^2\rho_D}\right)^2}\right]} \tag{2.12}$$

for the non-Lyman lines or the ratio

$$ratio = \frac{\frac{3}{2} \frac{mv^2 n_aq_a}{(Z-1)^2}\left(\frac{\pi^2}{4} - 1\right)}{\frac{1}{2}\left[1 - \frac{3}{2}\frac{Z^2(Z-1)^2}{n_a^4 m^2 v^2}\right] + \log\left[\sqrt{\frac{3}{2}\frac{Z\,v\,\rho_D}{n_a^2}}\sqrt{1 + \left(\frac{Z-1}{m\,v^2\rho_D}\right)^2}\right]} \tag{2.13}$$

for the Lyman lines.

Table 1 presents the values of the ratio from Equation (2.13) for several Lyman lines of He II at the temperature T = 8 eV and the electron density $N_e = 2 \times 10^{17}$ cm^{-3} [14].

Table 1. Ratio from Equation (2.13) for the Stark components of several Lyman lines of He II. at the temperature T = 8 eV and the electron density $N_e = 2 \times 10^{17}$ cm^{-3} [14].

n	$|q|$	*Ratio*
2	1	0.3261
3	1	0.3748
3	2	0.7496
4	1	0.5156
4	2	1.0311
4	3	1.5467

Figure 1 shows the ratio from Equation (2.13) versus the electron density N_e for the Stark components of the electric quantum number $|q| = 1$ of Lyman-alpha ($n = 2$), Lyman-beta ($n = 3$), and Lyman-gamma ($n = 4$) lines of He II at the temperature T = 8 eV.

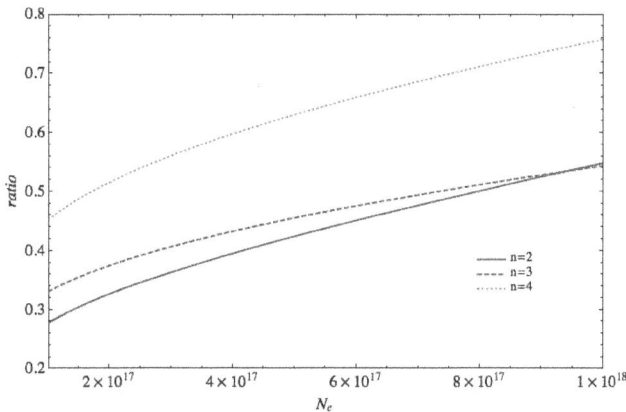

Figure 1. Ratio from Equation (2.13) versus the electron density N_e for the Stark components of the electric quantum number $|q| = 1$ of Lyman-alpha ($n = 2$), Lyman-beta ($n = 3$), and Lyman-gamma ($n = 4$) lines of He II at the temperature T = 8 eV [14].

It is seen that, for the electron broadening of the Lyman lines of He II, the allowance for the effect considered in [14] is indeed becomes significant already at electron densities $N_e \sim 10^{17}$ cm^{-3} and it increases with the growth of the electron density. The authors noted that when the ratio, formally calculated by Equation (2.13), becomes comparable to unity, this is the indication that the approximate analytical treatment based on expanding Equation (2.7) up to the first order of parameter β, is no longer valid. In this case, the calculations should be based on solving Equation (2.7) with respect to ρ without such approximation.

3. Center-of-Mass Effects for Hydrogen Atoms in a Nonuniform Electric Field: Applications to Magnetic Fusion, Radiofrequency Discharges, and Flare Stars

In the recent paper [17], there was studied whether or not the Center-of-Mass (CM) motion and the relative motion can be separated for hydrogen atoms in a *nonuniform* electric field. For hydrogenic atoms/ions in a uniform electric field, it was well-known that the CM and relative motions can be separated rigorously (exactly)—see, e.g., [18]. As for hydrogenic atoms/ions in a nonuniform electric field, there seemed to be nothing about the separation (or non-separation) of the CM and relative motions in the literature, to the best knowledge of the author of [17].

So, the author of [17] first treated a general case by considering a system of two charges e_1 and e_2 of masses m_1 and m_2, respectively, in a nonuniform electric field that is represented by the potential φ. After substituting

$$\mathbf{R} = (m_1\mathbf{r}_1 + m_2\mathbf{r}_2)/(m_1 + m_2), \tag{3.1}$$

$$\mathbf{r} = \mathbf{r}_2 - \mathbf{r}_1, \tag{3.2}$$

so that $\mathbf{R}$ and $\mathbf{r}$ are the coordinates that are related to the CM motion and the relative motion, and proceeding from the Lagrangian L to the Hamiltonian H, there was obtained the following expression for the latter [17]:

$$H = H_{CM}(\mathbf{R}, \mathbf{P}) + U(\mathbf{R}, \mathbf{r}) + H_r(\mathbf{r}, \mathbf{p}) \tag{3.3}$$

Here,

$$H_{CM}(\mathbf{R}, \mathbf{P}) = P^2/[2(m_1 + m_2)] + (e_1 + e_2)\varphi(\mathbf{R}) \tag{3.4}$$

is the Hamiltonian of the CM, $\mathbf{P}$ being the momentum of the CM motion;

$$H_r(\mathbf{r}, \mathbf{p}) = p^2/(2\mu) + e_1e_2/r \tag{3.5}$$

is the Hamiltonian of the relative motion, $\mathbf{p}$ being the momentum of the relative motion;

$$U(\mathbf{R}, \mathbf{r}) = \mu(e_1/m_1 - e_2/m_2)\mathbf{r}\mathbf{F}(\mathbf{R}) \tag{3.6}$$

is the coupling of the CM and relative motions. In Equation (3.6)

$$\mu = m_1m_2/(m_1 + m_2) \tag{3.7}$$

is the reduced mass of the two particles, and

$$\mathbf{F}(\mathbf{R}) = -d\varphi(\mathbf{R})/d\mathbf{R} \tag{3.8}$$

is a nonuniform electric field (in the expansion of the electric potential, the terms higher than the dipole one were disregarded). In Equation (3.6) and below, for any two vectors $\mathbf{A}$ and $\mathbf{B}$, the notation $\mathbf{AB}$ stands for the scalar product (also known as the dot-product) of the two vectors.

Thus, Equations (3.3) and (3.6) showed that at the presence of a nonuniform electric field, the CM motion and the relative motion are coupled (by $U(\mathbf{R}, \mathbf{r})$ from Equation (3.6)), and therefore, rigorously speaking, they cannot be separated. However, in the case where $m_1 << m_2$, the CM and relative motions can be separated by using the approximate analytical method of separating rapid and slow subsystems: in this case, the characteristic frequency of the relative motion is much greater than the characteristic frequency of the CM motion, so that the former and the latter are the rapid and slow subsystems, respectively. By applying this method (details of this method that can be found, e.g., in [15]) the author of [17] achieved the pseudoseparation of the CM motion and the relative motion and obtained the following expression for the effective Hamiltonian of the CM motion:

$$H_{CM,eff}(\mathbf{R}, \mathbf{P}) = P^2/[2(m_1 + m_2)] + (e_1 + e_2)\varphi(\mathbf{R}) - (3n|q|\hbar^2/2)[1/(m_1e_2) + 1/(m_2|e_1|)] \\ F(\mathbf{R})\cos[\theta(\mathbf{R})] \tag{3.9}$$

Here, $\theta(\mathbf{R})$ is the polar angle of the vector $\mathbf{F}(\mathbf{R})$, the z-axis being chosen along the unperturbed Runge-Lenz vector $\mathbf{A}$; q is the electric quantum number ($q = n_1 - n_2$, where n_1 and n_2 are the parabolic quantum numbers).

The author of [17] emphasized that he treated the CM coordinate $\mathbf{R}$ as the *dynamical variable* (which generally depends on time) and that the Hamiltonian $H_{CM,eff}(\mathbf{R}, \mathbf{P})$ from Equation (3.9) can be used to solve for the CM motion. This was the primary distinction of his work from papers where the CM coordinate of a hydrogenic atom/ion in a nonuniform electric fields was considered to be fixed.

In the particular case of hydrogen atoms, one has

$$e_1 = e, e_2 = -e, \mu = m_e m_p/(m_e + m_p) \tag{3.10}$$

where e > 0 is the electron charge, m_e and m_p are the electron and proton masses, respectively. Then, Equation (3.9) simplifies to [17]:

$$H_{CM,eff}(\mathbf{R}, \mathbf{P}) = P^2/(2m) - [3n|q|\hbar^2/(2\mu e)] F(\mathbf{R})\cos[\theta(\mathbf{R})], m = (m_e + m_p) \tag{3.11}$$

Next, the author of [17] considered the situation where the nonuniform electric field is due to the nearest (to the hydrogen atom) ion of the positive charge Ze and mass m_i in a plasma that is located at the distance $\mathbf{R}$ from the hydrogen atom. Then, the Hamiltonian from Equation (3.11) was rewritten as

$$H_{CM,eff}(\mathbf{R}, \mathbf{P}) = P^2/(2m) - (D/R^2)\cos\theta, D = [3n|q|\hbar^2/(2\mu)] Z, \cos\theta = \mathbf{AR}/AR \tag{3.12}$$

This Hamiltonian represents a particle of mass m in the dipole potential. Since this particle is relatively heavy ($m >> m_e$), its motion can be described classically and the corresponding classical solution is well-known—see, e.g., paper [19]. For this physical system, the radial motion can be exactly separated from the angular motion resulting in the following radial equation:

$$m[R(dR/dt) + (dR/dt)^2] = E_{CM} \tag{3.13}$$

where E_{CM} is the total energy of the particle. This equation allows for the following exact general solution:

$$R(t) = (2E_{CM}t^2/m + 2R_0v_0t + R_0^2)^{1/2}, R_0 = R(0), v_0 = (dR/dt)_{t=0} \tag{3.14}$$

It was well-known that in plasmas of relatively low electron densities N_e, the Stark broadening of the most intense hydrogen lines, i.e., the lines corresponding to the radiative transitions between the levels of the low principal quantum numbers (such as, e.g., Ly-alpha, Ly-beta, H-alpha, etc.), is dominated by the ion dynamical broadening—see, e.g., publications [5,20–25]. The corresponding validity condition is presented in Appendix of paper [17]. In the so-called "conventional theory" of the dynamical Stark broadening (also known as the "standard theory") [26–29], the relative motion within the pair "radiator—perturber" was assumed to occur along a straight line—as for a *free motion* (in our case the radiator is a hydrogen atom and the perturber is the perturbing ion).

However, from the preceding discussion in paper [17] it followed that in the more advanced approach, the relative motion within the pair "radiator-perturber" should be treated as the motion in the dipole potential—$(D/R^2)\cos\theta$, as seen from Equation (3.12). This approach modifies the cross-sections $\sigma(V_0)$ of so-called optical collisions, i.e., the cross-sections of collisions leading to virtual transitions inside level a between its sublevels and to virtual transitions inside level b between its sublevels, resulting in the broadening of Stark components of the hydrogen spectral line. The ion dynamical broadening operator Φ_{ab} is related to $\sigma(V_0)$, as follows

$$\Phi_{ab}(t) = -\int dV_0\, f(V_0)\, N_i\, V_0\, \sigma(V_0) \tag{3.15}$$

where V_0 is the relative velocity within the pair "radiator-perturber" at $t = 0$.

By considering the motion within the pair "radiator-perturber" in the reference frame where the perturbing ion is at rest, the authors of [17] obtained the following analytical expression for the matrix elements of the operator σ

$$_{\alpha\beta}(\sigma)_{\beta\alpha',A,D} = 2\pi\cdot{}_{\alpha\beta}(K^2)_{\beta\alpha}Q_0\{\ln[(\exp(2b^2)-1)^{1/2}(1/w^4-1)^{1/4}/2^{1/2}] - b^2/2 + [1/(4w^2)]\ln[(1+w^2)/(1-w^2)]\} \tag{3.16}$$

where Q_0

$$Q_0 = 2Z^2\hbar^2/(3\mu^2V_0{}^2) \tag{3.17}$$

and

$$_{\alpha\beta}(K^2)_{\beta\alpha} = (9/8)[n^2(n^2+q^2-m^2-1) - 4nqn'q' + n'^2(n'^2+q'^2-m'^2-1)] \tag{3.18}$$

and

$$b = [2C/(3Z)]^{1/2}(n^2+n'^2)^{1/2}/(n^2-n'^2) \tag{3.19}$$

and

$$w = [2e\hbar/(\mu T)][(n^2+n'^2)Zm_rN_e]^{1/2} = 8.99 \times 10^{-10}[(n^2+n'^2)ZN_em_r/m_p]^{1/2}/T \tag{3.20}$$

where

$$m_r = (m_e+m_p)m_i/(m_e+m_p+m_i) \tag{3.21}$$

and C is the so-called strong collision constant ($C \leq 2$). In the utmost right part of Equation (3.20), the temperature T is in eV and the electron density N_e is in cm^{-3}.

For presenting the effect of the CM motion in the universal form, it is convenient (as the authors of [17] did) to introduce the ratio of the cross-section $_{\alpha\beta}(\sigma)_{\beta\alpha,A,D}$ to the corresponding cross-section $_{\alpha\beta}(\sigma)_{\beta\alpha,G}$ from the conventional theory by Griem [29]. The ratio of the cross-sections is essentially the same as the ratio of widths $\gamma_{\alpha\beta,A,D}/\gamma_{\alpha\beta,G}$:

$$\text{ratio} = {}_{\alpha\beta}(\sigma)_{\beta\alpha,A,D}/{}_{\alpha\beta}(\sigma)_{\beta\alpha,G} = \gamma_{\alpha\beta',A,D}/\gamma_{\alpha\beta,G} = \{\ln[(\exp(2b^2)-1)^{1/2}(1/w^4-1)^{1/4}/2^{1/2}] - b^2/2 + [1/(4w^2)]\ln[(1+w^2)/(1-w^2)]\}/\{\ln[b/(wC^{1/2})] + 0.356\} \tag{3.22}$$

This ratio is a universal function of just two dimensionless parameters w and b that are applicable for any set of the five parameters N_e, T, n, n', and C.

The authors of [17] provided numerical examples for some laboratory and astrophysical plasmas where the allowance for the CM motion significantly affects the ion dynamical Stark width. The first example was edge plasmas of magnetic fusion machines (such as, e.g., tokamaks), characterized by the electron density $N_e = (10^{14}-10^{15})$ cm^{-3} and the temperature of one or few eV (see, e.g., review [30]). For these plasma parameters, the Stark broadening of the most intense hydrogen spectral lines (Ly-alpha, Ly-beta, H-alpha, etc.) can be dominated by the ion dynamical broadening (see, e.g., [5,20–25]).

The second example was plasmas in the atmospheres of flare stars. They are characterized by practically the same range of plasma parameters as the edge plasmas of magnetic fusion machines—see, e.g., book [31] and paper [32].

Figure 2 presents this ratio (for the H_α line emitted from a hydrogen plasma) versus the electron density N_e at T = 1 eV for C = 2 (solid line) and for C = 3/2 (dashed line). It is seen that the allowance for the CM motion increases the ion dynamical Stark width of the H_α line in these kinds of plasmas by up to (15–20)%.

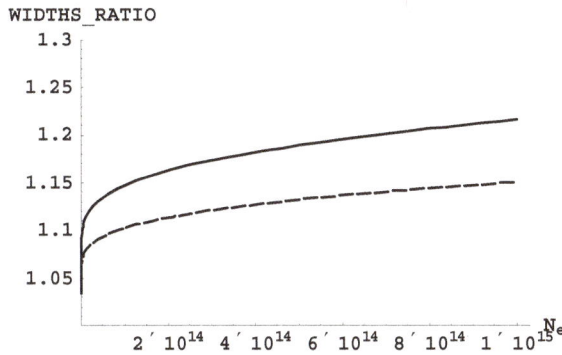

Figure 2. The ratio of the ion dynamical Stark width with the allowance for the center-of-mass motion to the ion dynamical Stark width from the conventional theory [29] versus the electron density N_e (cm^{-3}) for the H_α line emitted from a hydrogen plasma at T = 1 eV for C = 2 (solid line) and for C = 3/2 (dashed line) [17]. Plasma parameters correspond to edge plasmas of magnetic fusion machines and to atmospheres of flare stars.

Another example from [17] relates to the plasmas of radiofrequency discharges, such as, e.g., those studied in papers [33–35]. The plasma parameters, e.g., in the experiments [33,34], are $N_e = 1.2 \times 10^{13}$ cm^{-3} and T = (1850–2000) K, i.e., T = (0.16–0.17) eV. Figure 3 presents this ratio (for the H_α line emitted from a hydrogen plasma) versus the electron density N_e at T = 0.17 eV for C = 2 (solid line) and for C = 3/2 (dashed line). It is seen that the allowance for the CM motion increases the ion dynamical Stark width of the H_α line in these kinds of plasmas by up to (15–20)%.

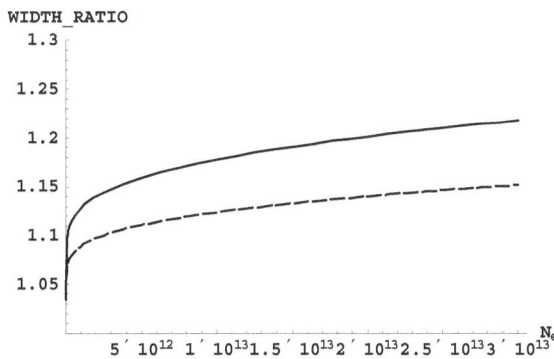

Figure 3. The ratio of the ion dynamical Stark width with the allowance for the center-of-mass motion to the ion dynamical Stark width from the conventional theory [29] versus the electron density N_e (cm^{-3}) for the H_α line emitted from a hydrogen plasma at T = 0.17 eV for C = 2 (solid line) and for C = 3/2 (dashed line) [17]. Plasma parameters correspond to radiofrequency discharges.

Thus, in addition to the fundamental importance, the results paper [17] seem to also have practical importance for spectroscopic diagnostics of laboratory and astrophysical plasmas.

4. New Source of Shift of Hydrogenic Spectral Lines in Plasmas: Analytical Treatment of the Effect of Penetrating Ions

4.1. Preamble

Red shifts of spectral lines (hereafter, SL) play an important role in astrophysics. Indeed, the relativistic (cosmological and gravitational) red shifts (see, e.g., book by Nussbaumer and Bieri [36]) are at the core of models of the Universe and of tests for the general relativity. However, for inferring the relativistic red shifts from the observed red shifts it is required to allow for the Stark shift of SL. Hydrogen and hydrogenlike (hereafter, H-like) SL in plasmas are usually shifted to the red by electric microfields—see, e.g., books by Griem [2] and by Oks [5] and references therein. Besides, in laboratory plasmas, measurements of the Stark shift can supplement measurements of the Stark width and thus enhance the plasma diagnostics—specifically the determination of the electron density (see, e.g., paper by Parigger et al. [37]).

In papers [38,39], there was described a new source of the Stark shift of hydrogenic SL—in addition to the previously known sources of the shift (we call the latter "standard shifts"). The new source of shift is due to configurations where the perturbing ion is *within* the bound electron cloud ("penetrating configurations"). The contribution to the shift from penetrating configurations is a product of two factors. The first fact is the statistical weight of penetrating configurations, which is relatively small. The second factor—the shift relevant to penetrating configurations—is relatively large. In papers [38,39] it was shown that the product of these two factors could exceed (sometimes, very significantly) the total standard shift.

In paper [38] the focus was at highly-excited (high-n) hydrogen SL, such as, e.g., Balmer lines of $n = 13$–17, studied in astrophysical and laboratory observations at the electron density $N_e \sim 10^{13}$ cm^{-3} by Bengtson and Chester [34]. Specifically, the authors of [34] presented the red shifts of these SL observed in the spectra from Sirius and in the spectra from a radiofrequency discharge plasma in the laboratory: both types of the observations yielded red shifts that exceeded the corresponding "standard" theoretical shifts by orders of magnitude. In paper [38], it was shown that for the high-n hydrogen lines, the contribution to the red shift from penetrating configurations is by orders of magnitude greater than the standard theoretical and that the allowance for this additional red shift removes the existed huge discrepancy between the observed and theoretical shifts of those high-n hydrogen SL.

In paper [39], the focus was on the contribution to the shift from penetrating configurations for hydrogenlike (H-like) SL. As an example, the authors of [39] compared their theoretical results with the experimental shift of the Balmer-alpha SL of He II 1640 A measured in a laboratory plasma by Pittman and Fleurier [40]. It was shown in [39] that the allowance for this new additional red shift leads to a good agreement with the measured shift from [40] for the entire range of the electron density being employed in that experiment, while without this new shift the standard shifts underestimated the measured shift by factors between two and five.

Below, we outline the theory behind the contribution to the shift from penetrating configurations—both for hydrogen SL by following paper [38], and for H-like SL by following paper [39].

4.2. "Standard" Shifts of Hydrogenic Spectral Lines

One of the most significant "standard" contributions to the shift of H-like SL is caused by quenching, non-zero Δn (Griem paper [41]), and elastic, zero Δn (Boercker and Iglesias paper [42]) collisions with plasma electrons—hereafter, the electronic shift (see also Griem paper [43]).

For H-like lines, one should also take into account a so-called plasma polarization shift (PPS). It plays an important role in explaining the observed shifts of the high-n H-like SL—see, e.g., books by Griem [2] and by Salzman [9] and paper by Renner et al. [44]. The PPS is less significant for the low-n H-like SL. Physically, the PPS is caused by the redistribution of plasma electrons due to the attraction to the radiating ion. When only plasma electrons inside the orbit of the bound electron were taken into account, the resulting theoretical PPS was blue (such as, e.g., in paper by Berg et al. [45]). Later, it was found that after the allowance for redistributed plasma electrons both outside and inside the bound electron orbit, the resulting theoretical PPS becomes red. However, theoretical results for red PPS by different authors differ by a factor of two—more details and the references will be provided below, while comparing the theoretical and experimental results.

Then, there is a controversial issue of the "standard' shift caused by plasma ions-hereafter, the standard ionic shift. Various existing calculations were based on the multipole expansion with respect to the ratio r_{rms}/R (in the binary description of the ion microfield) or with respect to the analogous parameter $r_{rms}F^{1/2}$ (in the multi-particle description of the ion microfield F). Here, r_{rms} is the root-mean-square value of the radius-vector of the atomic electron ($r_{rms} \sim n^2/Z_1$, where Z_1 is the nuclear charge), and R is the separation between the nucleus of the radiating atom/ion and the nearest perturbing ion. We use the atomic units here and below.

The dipole term of the expansion ($\sim 1/R^2$ or $\sim F$) does not lead to any shift of a hydrogenic SL. Indeed, each pair of the Stark components, as characterized by the electric quantum numbers q and $-q$, is symmetric with respect to the unperturbed frequency ω_0 of the hydrogenic line—symmetric concerning both the displacement from ω_0 and the intensity. Here, $q = n_1 - n_2$, where n_1 and n_2 are the first two of the three parabolic quantum numbers ($n_1 n_2$ m). The next, quadrupole term of the expansion ($\sim 1/R^3$ or $\sim F^{3/2}$) does not shift the center of gravity of hydrogenic lines. This was rigorously proven in paper [46]. Namely, after allowing for the quadrupole corrections to both the energies/frequencies and the intensities, and then summing up over all the Stark components of a hydrogenic SL, the center of gravity shift becomes exactly zero at any fixed value of R or F.

Thus, within the approach based on the multipole expansion, the first non-vanishing ionic contribution to the shift of hydrogenic SL is supposed to originate from the next term of the multipole expansion: from the term $\sim 1/R^2$ or $\sim F^2$. In processing this term, many authors considered only the quadratic Stark (QS) effect—see papers by Griem [43] and by Könies and Günter [47,48]:

$$\Delta E_{QS}{}^{(4)} = -\frac{Z_2{}^2 n^4}{16\, Z_1{}^4 R^4}\left(17n^2 - 3q^2 - 9m^2 + 19\right) \tag{4.1}$$

Here, Z_2 is the charge of perturbing ions; the superscript (4) at ΔE_{QS} specifies that this term is of the fourth order with respect to the small parameter r_{rms}/R.

However, first, the corrections of this order to the energies are of the same order as the corrections to the intensities, as noted in paper by Demura et al. [49]. Therefore, calculations in Könies and Günter papers [47,48] were inconsistent because they took into account the quadratic Stark corrections only to energies.

Second, there is an even more important flaw in papers by Griem [43] and by Könies and Günter [47,48], as follows. The above Equation (4.1) was obtained by using the dipole term of the multipole expansion treated in the second order of the perturbation theory. However, the quadrupole term, processed in the second order of the perturbation theory, and the octupole term, processed in the 1st order of the perturbation theory, in fact also yield energy corrections $\sim 1/R^4$—this was shown as early as in 1969 by Sholin [50]. The rigorous energy correction of the order $\sim 1/R^4$ can be obtained in the form (given in Sholin paper [50] and presented also in book by Komarov et al. [51]):

$$\Delta E^{(4)} = \frac{Z_2 n^3}{16 Z_1{}^4 R^4}\left[Z_1 q\left(109q^2 - 39n^2 - 9m^2 + 59\right) - Z_2 n\left(17n^2 - 3q^2 - 9m^2 + 19\right)\right] \tag{4.2}$$

Apparently, it is inconsistent to allow for one term and to neglect two other terms of the same order of magnitude.

Nevertheless, from table III of Griem paper [43] it is clear that the ionic shift $\Delta E^{(4)}$ due to the quadratic Stark effect is by one or more orders of magnitude smaller than the corresponding electronic shift (and that while the latter is red, the former is blue). A more consistent calculation of the ionic shift $\Delta E^{(4)}$ does not change the fact that it is just a very small correction to the corresponding electronic shift and is even a smaller correction to the sum of the corresponding electronic shift and the PPS.

An intermediate summary: the standard shift can be represented with the accuracy, sufficient for comparison with experiments/observations, by the electronic shift for hydrogen SL, or by the sum of the electronic shift and the PPS for H-like SL, while the standard ionic shift can be neglected.

Table 2 presents the electronic shift S_e of the hydrogen SL H_{13}–H_{17}, calculated by formulas from papers by Griem [41,43], and their comparison with the shifts from paper by Bengtson & Chester [34] observed in astrophysical and laboratory plasmas, the latter being a radofrequency discharge of $N_e \sim 10^{13}$ cm^{-3}. It is seen that the electronic shift is by orders of magnitude smaller than both the shift of the SL H_{14}, H_{15}, H_{17} observed in the spectrum of Sirius and the shift of the SL H_{13}, H_{15}, and H_{17} observed in the laboratory plasma.

Table 2. Electronic shift S_e of the hydrogen spectral lines H_{13}–H_{17}, as calculated by formulas from papers by Griem [41,43] and their comparison with the shifts from paper by Bengtson and Chester [34] observed in astrophysical (S_{Sirius}) and laboratory (S_{exp}) plasmas. All of the shifts are in Angstrom.

n	λ_n(A)	S_e(A)	S_{Sirius}(A)	S_{exp}(A)
13	3734	0.0017		0.03 ± 0.03
14	3722	0.0021	0.03 ± 0.05	0.00 ± 0.04
15	3712	0.0026	0.09 ± 0.07	0.15 ± 0.05
16	3704	0.0032	-0.007 ± 0.05	0.00 ± 0.05
17	3697	0.0038	0.21 ± 0.08	0.30 ± 0.08

4.3. New Source of the Red Shift and the Comparison with Experiments/Observations

The standard approaches to calculating the ionic contribution to the shift of hydrogenic SL, discussed in the previous section, used the multipole expansion in terms of the parameter r_{rms}/R that was considered to be small. All the terms of the multipole expansion, starting from the quadrupole term, at the averaging over the distribution of the separation R between the nucleus of the radiating atom/ion and the nearest perturbing ion, led to integrals diverging at small R. These diverging integrals were evaluated one way or another, e.g., by introducing cutoffs. However, the mere fact that the integrals were diverging was an indication that the standard approach did not provide a consistent complete description of the ionic contribution to the shift.

The fact is that the standard approaches disregarded configurations where $r_{rms}/R > 1$, i.e., where the nearest perturbing ion is within the radiating atom/ion (below we call them "penetrating configurations"). The contribution to the ionic shift from penetrating configurations is a product of two factors. The first fact is the statistical weight of penetrating configurations, which is relatively small. The second factor—the shift relevant to penetrating configurations—is relatively large. In papers [38,39] it was shown that the product of these two factors could exceed the total standard shift.

For penetrating configurations, it is appropriate to use the expansion in terms of the parameter $R/r_{rms} < 1$ in the basis of the spherical wave functions of the so-called "united atom", which is a hydrogenic ion of the nuclear charge $Z_1 + Z_2$. The energy expansion has the form (see, e.g., book by Komarov et al. [51], Equations (5.10)–(5.12)):

$$E = -(Z_1 + Z_2)^2/(2n^2) + O(R^2/r_{rms}^2) \tag{4.3}$$

Therefore, the first non-vanishing contribution to the shift of the energy level is

$$S(n) = -(Z_1 + Z_2)^2/(2n^2) - [-Z_1^2/(2n^2)] = -(2Z_1Z_2 + Z_2^2)/(2n^2) \qquad (4.4)$$

While S(n) scales $\sim 1/n^2$, the statistical weight I(n) increases with growing n more rapidly than $\sim n^2$, as shown in [38,39]. Therefore, the sign of the shift of the spectral line is determined by the sign of the shift of the upper level: it is negative in the frequency scale, so that it is positive in the wavelength scale—the red shift.

The final step is the integration over the distribution of the interionic distances, which can be obtained from the distribution of the ion microfield that is presented in papers by Held [52] and Held et al. [53], where these authors took into account ion-ion correlations and the screening by plasma electrons. The upper limit of the integration could be taken as the root mean square size of the bound electron cloud.

The results of calculating this contribution to the shift of the hydrogen SL H_{13}–H_{17} by for the parameters corresponding to the observations from by Bengtson and Chester [34] ($N_e = 1.2 \times 10^{13}$ cm^{-3}, $Z_1 = Z_2 = 1$), are shown in Table 3 in the column $S_{i,penetr}$—according to paper [39]. The sum $S_{i,penetr} + S_e$ is shown in the column S_{tot}. The theoretical error margin is shown only for the latter and it is primarily due to the approximate way of estimating $S_{i,penetr}$.

Table 3. Shift $S_{i,penetr}$ due to penetrating ions [39] and its sum S_{tot} with the electronic shift S_e for the hydrogen spectral lines H_{13}–H_{17}, and the comparison with the shifts from paper by Bengtson and Chester [34] observed in astrophysical (S_{Sirius}) and laboratory (S_{exp}) plasmas. All of the shifts are in Angstrom.

n	λ_n(A)	S_e(A)	$S_{i,penetr}$(A)	S_{tot}(A)	S_{Sirius}(A)	S_{exp}(A)
13	3734	0.0017	0.032	0.034 ± 0.010		0.03 ± 0.03
14	3722	0.0021	0.043	0.045 ± 0.014	0.03 ± 0.05	0.00 ± 0.04
15	3712	0.0026	0.057	0.060 ± 0.018	0.09 ± 0.07	0.15 ± 0.05
16	3704	0.0032	0.073	0.076 ± 0.023	-0.007 ± 0.05	0.00 ± 0.05
17	3697	0.0038	0.093	0.10 ± 0.03	0.21 ± 0.08	0.30 ± 0.08

The following can be seen from Table 3.

For the SL H_{13}, there is an excellent agreement between the total theoretical shift S_{tot} and the experimental shift S_{exp}. No data for the shift of this SL from Sirius.

For the SL H_{14}, there is a good agreement of the total theoretical shift with the shift of this SL observed from Sirius and a satisfactory agreement (within the error margins) with the experimental shift of this SL.

For the SL H_{15}, there is a good agreement of the total theoretical shift with the shift of this SL observed from Sirius and a satisfactory agreement (almost within the error margins) with the experimental shift of this SL.

For the SL H_{16}, there is a satisfactory agreement (within the error margins) of the total theoretical shift with both the shift of this SL as observed from Sirius and the experimental shift of this SL.

For the SL H_{17}, there is a satisfactory agreement (within the error margins) of the total theoretical shift with the shift of this SL observed from Sirius, but a disagreement with the experimental shift of this SL; however, the latter disagreement is not anymore by two orders of magnitude, as it was the case before the allowance for the shift by penetrating ions, but rather just by a factor of two (after allowing for the error margins).

As another example taken from paper [39], here is the comparison of various theoretical sources of the shift (including the shift by penetrating ions) for the He II Balmer-α line with the experimental shift of this line that was obtained by Pittman and Fleurier [40] for the electron densities in the range of $N_e = (0.3 - 2.3) \times 10^{17}$ cm^{-3}. In Figure 4, the experimental shifts $\Delta\lambda_{exp}$ are shown by circles. The theoretical shift by Griem [41,43] $\Delta\lambda_{Griem}$, with which Pittman and Fleurier [40] compared their experimental results, is shown by the dashed blue line.

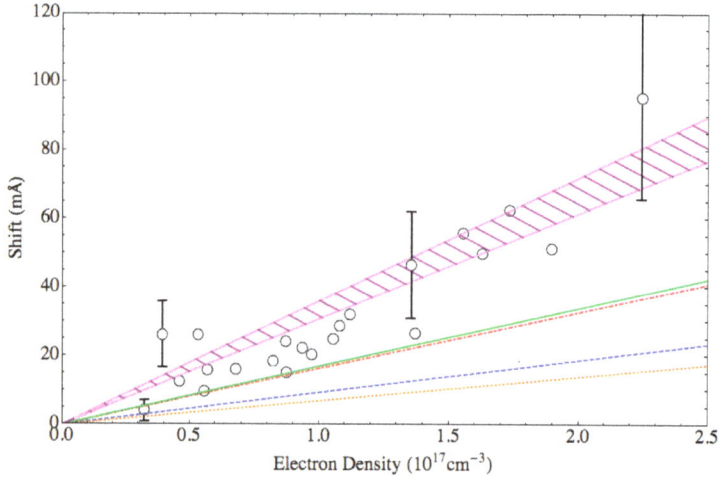

Figure 4. Comparison of the experimental shift of the He II Balmer-alpha line 1640 A measured by Pittman and Fleurier [40], shown by circles, with the following theoretical shifts: Griem's shift [41,43]—dashed blue line; plasma polarization shift-dotted orange line; the sum of the latter two theoretical shifts-dashed-dotted red line; shift due to penetrating ions [39]—solid green line; the sum of all three theoretical shifts-purple band, the width of which reflects the theoretical error. The experimental error bars are shown only for few electron densities in order to avoid making the figure too "busy" and difficult to understand.

It is seen that there was a huge discrepancy between the experimental red shift $\Delta\lambda_{exp}$ and the theoretical red shift by Griem. The discrepancy is by a factor of 2.6 at $N_e = 10^{17}$ cm^{-3} and is increasing to almost a factor of five at $N_e = 2.2 \times 10^{17}$ cm^{-3}.

In Figure 4, the PPS $\Delta\lambda_{PPS}$ is shown by the dotted red line. The sum $\Delta\lambda_{Grie} + \Delta\lambda_{PPS}$ is shown by the dash-dotted brown line. It is seen that even after adding the PPS to Griem's shift, their sum still underestimates the experimental shift at least by a factor of two.

The shift due to penetrating ions $\Delta\lambda_{PI}$ is shown by the solid green line. The sum $\Delta\lambda_{Griem} + \Delta\lambda_{PPS} + \Delta\lambda_{PI}$ is presented in Figure 4 by the dashed purple band (the width of the band reflects the theoretical error of this sum, originated from the relative inaccuracy of the relatively simple model from [39] and from the theoretical uncertainty of the PPS). It is seen that adding the shift due to the penetrating ions brings the total shift into a good agreement with the experimental shift.

5. Revision of the Method for Measuring the Electron Density Based on the Asymmetry of Hydrogenic Spectral Lines in Dense Plasmas

In paper [54], there was previously proposed and experimentally implemented a new diagnostic method for measuring the electron density N_e while using the asymmetry of hydrogenic spectral lines in dense plasmas. In that paper, in particular, from the experimental asymmetry of the C VI Lyman-delta line that was emitted by a vacuum spark discharge, the electron density was deduced to be $N_e = 3 \times 10^{20}$ cm^{-3}. This value of N_e was in good agreement with the electron density that was determined from the experimental widths of C VI Lyman-beta and Lyman-delta lines.

Later, this diagnostic method was employed also in the experiment presented in paper [55]. In that laser-induced breakdown spectroscopy experiment, the electron density $N_e \sim 3 \times 10^{17}$ cm^{-3} was determined from the experimental asymmetry of the H I Balmer-beta (H-beta) line.

When compared to the traditional method of deducing N_e from the experimental widths of spectral lines, the new method has the following advantages. First, the traditional method requires measuring widths of at least two spectral lines (to isolate the Stark broadening from competing broadening mechanisms), while for the new diagnostic method it is sufficient to obtain the experimental profile of just one spectral line. Second, the traditional method would be difficult to implement if the center of the spectral lines is optically thick, while the new diagnostic method can still be used even in this case.

In the theory underlying this new diagnostic method, the contribution of plasma ions to the spectral line asymmetry was calculated only for configurations where the perturbing ions are outside the bound electron cloud of the radiating atom/ion (non-penetrating configurations). In paper [56], the authors took into account the contribution to the spectral line asymmetry from *penetrating configurations*, where the perturbing ion is inside the bound electron cloud of the radiating atom/ion. While calculating the corresponding corrections to the wave functions and to the intensity of spectral line components, they employed the robust perturbation theory developed in paper [57].

The theory from paper [57], which is applicable to degenerate quantum systems, constructs the perturbation approach to the operator of an additional conserved quantity (rather than to the Hamiltonian operator). This theory avoids infinite summations that are encountered in the standard perturbation theory. For the problem considered in paper [56], the additional conserved quantity was the super-generalized Runge-Lenz vector in the two-Coulomb center problem [58].

Using this advanced approach, the authors of paper [56] showed analytically that, in high density plasmas, the allowance for penetrating ions can result in significant corrections to the electron density deduced from the spectral line asymmetry. Since paper [56] is published in the same Special Issue as this review, here we would only outline very briefly the main result of paper [56].

Table 4 (which reproduces Table 1 from [56]) presents the following quantities for the He II Balmer-alpha line at five different values of the actual electron density:

- the theoretical degree of asymmetry ρ_{act} calculated with the allowance for penetrating ions,
- the theoretical degree of asymmetry ρ_{quad} calculated without the allowance for penetrating ions,
- the electron density $N_{e,quad}$ that would be deduced from the experimental asymmetry degree while disregarding the contribution of the penetrating ions, and
- the relative error $|N_{e,quad} - N_{e,act}|/N_{e,act}$ in determining the electron density from the experimental asymmetry degree while disregarding the contribution of the penetrating ions.

Table 4. The relative error in determining the electron density N_e from the experimental asymmetry degree while disregarding the contribution of the penetrating ions for the He II Balmer-alpha line. The physical quantities in Table 4 are explained in the text directly above Table 4.

$N_{e,act}/(10^{18}$cm$^{-3})$	ρ_{act}	ρ_{quad}	$N_{e,quad}/(10^{18}$cm$^{-3})$	$|N_{e,quad} - N_{e,act}|/N_{e,act}$
2	0.0925	0.0955	1.82	9.03%
4	0.114	0.120	3.42	14.5%
6	0.128	0.138	4.86	19.1%
8	0.139	0.152	6.16	23.1%
10	0.147	0.163	7.33	26.7%

It is seen that in high density plasmas, the allowance for penetrating ions can indeed result in significant corrections to the electron density deduced from the spectral line asymmetry.

6. Lorentz–Doppler Broadening of Hydrogen/Deuterium Spectral Lines: Analytical Solution for Any Angle of Observation and Any Magnetic Field Strength, and Its Applications to Magnetic Fusion and Solar Physics

6.1. Preamble

Strongly-magnetized plasmas are encountered both in astrophysics (e.g., in Sun spots, in the vicinity of white dwarfs etc.) and in laboratory plasmas (e.g., in magnetic fusion devices). In such plasmas, as hydrogen/deuterium atoms move across the magnetic field **B** with the velocity **v**, they experience a Lorentz electric field $E_L = v \times B/c$ in addition to other electric fields. The Lorenz field has a distribution because the atomic velocity **v** has a distribution. So, for radiating hydrogen/deuterium atoms this becomes an additional source of the *broadening* of spectral lines.

In paper [59], there were described situations where the Lorentz broadening serves as the primary broadening mechanism of Highly-excited Hydrogen/deuterium Spectral Lines (HHSL). One example that is discussed in paper [59] was HHSL that was emitted from edge plasmas of tokamaks. In laboratory plasmas, HHSL are used for measuring the electron density at the edge plasmas of tokamaks (see, e.g., papers [60,61] and Section 4.3 of review [62]) and in radiofrequency discharges (see, e.g., paper [63] and book [5]).

Another example that is discussed in paper [59] was HHSL emitted from the solar chromosphere. They are observed and used for measuring the electron density in the solar chromosphere (see, e.g., paper [64]).

One of the most interesting features of these situations is that the combination of Lorentz and Doppler broadenings cannot be taken into account simply as a convolution of these two broadening mechanisms, as it was pointed out for the first time in paper [65]. The Lorentz and Doppler broadening intertwine in a more complicated way. Indeed, let us consider a Stark component of HHSL. Its Lorentz–Doppler profile in the frequency scale is proportional (in the laboratory reference frame) to $\delta[\Delta\omega - (\omega_0 v/c)\cos\alpha - (kX_{\alpha\beta}Bv/c)\sin\vartheta]$, where in the argument of this δ-function the quantity α is the angle between the direction of observation and the atomic velocity **v**, and ϑ is the angle between vectors **v** and **B**.

In paper [59], there was derived a general expression for the Lorentz–Doppler profiles of HHSL for the arbitrary strength of the magnetic field **B** and for the arbitrary angle of the observation ψ with respect **B**. More *specific* analytical results were obtained in paper [57], only for $\psi = 0$ and $\psi = 90$ degrees. It was shown that a relatively strong magnetic field causes a significant *suppression of π-components* when compared to σ-components for the observation at $\psi = 90$ degrees, which was a counterintuitive result.[1]

In the subsequent paper [66], the authors obtain *specific* analytical results for the Lorentz–Doppler profiles of HHSL for the arbitrary strength of the magnetic field **B** and for an *arbitrary angle of the observation* ψ. In particular, it was shown in [66] that the effect of the suppression of π-components at a relatively strong magnetic field rapidly diminishes as the angle of observation ψ decreases from 90 degrees. Another finding in [66] was that the width of the Lorentz–Doppler profiles is a non-monotonic function of the magnetic field for observations perpendicular to **B**, which was yet another counterintuitive result. So, the results presented below should be important for spectroscopic diagnostics of magnetic fusion plasmas [67] and for solar plasmas.

----

[1] We note in passing that in paper [59] there were minor typographic errors in Equations (31) and (32). In Equation (31), the factor in front of the integral should be $\pi^{-1}|2w|^{-\frac{1}{2}}$. In Equation (32), the factor in front of the last brackets should be $[\Gamma(1/4)\Gamma(-1/4)]^{-1}|w|^{-1/2}$.

6.2. Analytical Results

For an arbitrary angle ψ between the direction of observation and the magnetic field, the relative configuration of vectors **B**, **E**$_L$, and **v**, as well as the choice of the reference frame is shown in Figure 5.

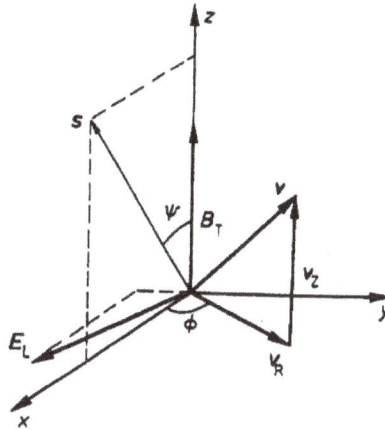

Figure 5. Relative configuration of the magnetic **B** and Lorentz **E**$_L$ fields and of the direction of the observation **s** ("s" stands for "spectrometer"). The z axis is along **B**. The direction of the observation **s** constitutes a non-zero angle ψ with **B**. The xz plane is spanned on vectors **B** and **s**. The atomic velocity **v** has a component v_z along **B** and a component v_R perpendicular to **B**. The component v_R constitutes an angle φ with the x axis [66].

In paper [59], for obtaining universal analytical results the following dimensionless notations were introduced:

$$w = c\,\Delta\omega/v_T\omega_0 = c\,\Delta\lambda/v_T\lambda_0, \quad b = kX_{\alpha\beta}B/\omega_0, \quad \mathbf{u} = \mathbf{v}/v_T \tag{6.1}$$

Here, w is the scaled detuning from the unperturbed frequency ω_0 or from the unperturbed wavelength λ_0 of a hydrogen spectral line, b is the scaled magnetic field, and **u** is the atomic velocity scaled with respect to the atomic thermal velocity v_T. The quantities k and $X_{\alpha\beta}$ in Equation (6.1) are

$$k = 3\hbar/(2m_e), \quad X_{\alpha\beta} = n_\alpha(n_1 - n_2)_\alpha - n_\beta(n_1 - n_2)_\beta \tag{6.2}$$

where n_1, n_2 are the parabolic quantum numbers, and n is the principal quantum numbers of the upper (subscript α) and lower (subscript β) Stark sublevels involved in the radiative transition.

A general expression for the Lorentz–Doppler profiles of components of HHSL for the arbitrary strength of the magnetic field **B** and for the arbitrary angle of the observation ψ with respect **B** was derived in paper [59], in the form of the following triple integral

$$I(w, b, \psi) = \int_0^\infty du_z f_z(u_z) \int_0^\infty du_R f_R(u_R) \int_0^\pi (d\phi/\pi) g(\psi, \phi)\, \delta[w - u_z\cos\psi - u_R(b + \sin\psi\,\cos\phi)] \tag{6.3}$$

where

$$f_z(u_z) = \frac{1}{\sqrt{\pi}}e^{-u_z^2}, \quad f_R(u_R) = 2\,u_R e^{-u_R^2}, \quad 0 < \psi < \frac{\pi}{2} \tag{6.4}$$

and g(ψ, φ) are factors different for π- and σ-components:

$$g_\pi(\psi) = 1 - \sin^2\psi\,\sin^2\phi, \quad g_\sigma(\psi) = \frac{1}{2}(1 + \sin^2\psi\,\sin^2\phi) \tag{6.5}$$

We note that the functions $f_z(u_z)$ and $f_R(u_R)$ are, respectively, the one-dimensional and the two-dimensional Maxwell distributions of the scaled atomic velocity $\mathbf{u} = \mathbf{v}/v_T$.

In paper [66], all three integrations in Equation (6.3) were performed analytically. The result is:

$$I(w, b, \psi) = \frac{e^{-\frac{w^2}{\cos^2 \psi}}}{\pi^{\frac{3}{2}} \cos \psi} \int_0^\pi \frac{g(\psi, \phi)}{|a(b,\psi,\phi)|^{\frac{3}{2}}} \left\{ \sqrt{a(b,\psi,\phi)} + \sqrt{\pi}\, c(w,b,\psi,\phi)\, e^{\frac{c(w,b,\psi,\phi)^2}{a(b,\psi,\phi)}} \left[1 + \text{Erf} \frac{c(w,b,\psi,\phi)}{\sqrt{a(b,\psi,\phi)}} \right] \right\} d\phi \quad (6.6)$$

where

$$a(b, \psi, \phi) = 1 + \frac{(b + \sin \psi \cos \phi)^2}{\cos^2 \psi}, \quad c(w,b,\psi,\phi) = w \frac{b + \sin \psi \cos \phi}{\cos^2 \psi} \quad (6.7)$$

In paper [66], the authors presented five different figures the Lorentz–Doppler profiles of π-components and another five different figures the Lorentz–Doppler profiles of σ-components of HHSL as calculated by Equations (6.6) and (6.7). Each figure shows profiles for five values of the angle ψ (in degrees): 0, 20, 45, 70, and 90. The figures differed from each other by the value of the scaled magnetic field b (defined in Equation (6.1)): b = 0.2, 0.5, 1, 2, and 5.

As an example, we reproduce below only two out of those ten figures. Namely, in Figures 6 and 7, the Lorentz–Doppler of π- and σ-components, respectively, are presented for b = 0.2.

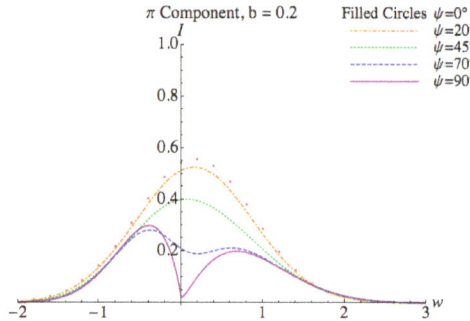

Figure 6. Lorentz–Doppler profiles of π-components of highly-excited hydrogen/deuterium spectral lines calculated by Equations (6.6) and (6.7), for the scaled magnetic field b = 0.2 (defined in Equation (6.1)) at five different values of the angle of observation ψ with respect to the magnetic field [66].

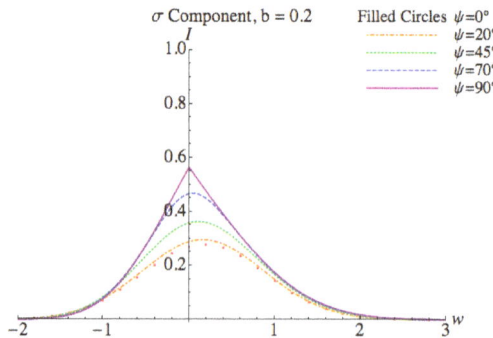

Figure 7. Lorentz–Doppler profiles of σ-components of highly-excited hydrogen/deuterium spectral lines calculated by Equations (6.6) and (6.7), for the scaled magnetic field b = 0.2 (defined in Equation (6.1)) at five different values of the angle of observation ψ with respect to the magnetic field [66].

From Figures 6 and 7 it is seen that, as the angle of the observation ψ increases from 0 to 90 degrees, the *π-components are suppressed*, while the σ-components are not; actually, the σ-components even becomes slightly more intense as ψ increases from 0 to 90 degrees. The suppression of the π-components is an *important counterintuitive result*.

Another interesting result from paper [66] is the following. The width of the Lorentz–Doppler profiles is a *non-monotonic function* of the scaled magnetic field b for observations perpendicular to **B**. As |b| increases from zero, the width first decreases, then reaches a minimum at |b| = 1 (i.e., when the shift in the Lorentz field is equal to the Doppler shift), and then increases—as presented in Figure 8 while using the Ly-beta line as an example. This is yet another *counterintuitive result*.

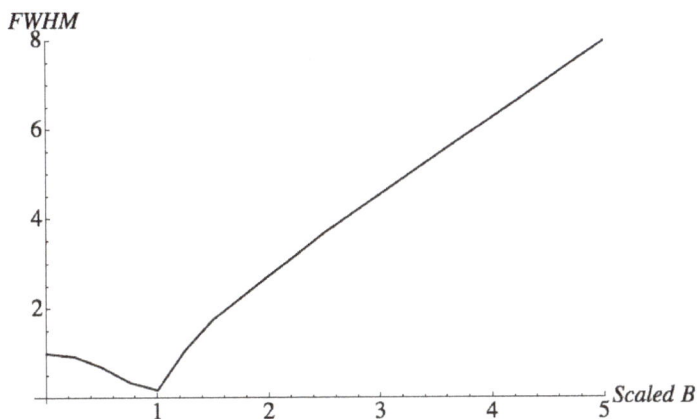

Figure 8. The Full Width at Half Maximum (FWHM) of the theoretical profile of hydrogen/deuterium Ly-beta line observed perpendicular to the magnetic field **B** with a polarizer along **B**. The scaled magnetic field is the ratio of the Lorentz-field shift to the Doppler shift. The FWHM is in units of the Doppler half width at half maximum. The narrowing effect is the most pronounced when the Lorentz-field shift is equal to the Doppler shift [66].

The decreasing part of the FWHM dependence on the magnetic field corresponds to relatively small magnetic fields: |b| < 1. In this range of |b|, the line profile has the bell shape. In this range of |b|, the complicated entanglement of the Doppler and Lorentz-filed mechanisms (that cannot be described as their convolution) causes the FWHM to decrease as |b| increases. This narrowing effect has a limited analogy with the well-known Dicke narrowing. Namely, in the Dicke case, the correlations between the Doppler mechanism and collisions cause the narrowing, while in our case, the correlations (the complicated entanglement) between the Doppler mechanism and the Lorentz-field mechanisms cause the narrowing. At relatively large magnetic fields, where |b| > 1, the line profile has the two-peak shape (one in the red part of the symmetric profile, another in the blue part of the symmetric profile). In this range of |b|, the Lorentz-field mechanism dominates over the Doppler mechanism. Therefore, as |b| increases in this range, the two peaks of the profile move further apart and the FWNM increases.

6.3. Validity and Applications

The above results become practically important when one can neglect the Stark broadening by the ion microfield and the Zeeman splitting. Here, are the corresponding validity conditions presented in paper [59].

The average Lorentz field

$$E_{LT} = Bv_T/c = 4.28 \times 10^{-3} \, B[T(K)]^{1/2} \tag{6.8}$$

$(v_T = (2T/M)^{1/2}$ is the atomic thermal velocity) can exceed the most probable ion microfield E_i when the magnetic field B exceeds the following critical value:

$$B_c = 4.69 \times 10^{-7} N_e^{2/3}/[T(K)]^{1/2} \qquad (6.9)$$

(in Equations (6.8) and (6.9), B is in Tesla). For example, in the solar chromosphere the typical plasma parameters are $N_e \sim 10^{11}$ cm^{-3} and $T \sim 10^4$ K (except solar flares where N_e can be higher by two orders of magnitude)—see, e.g., [64,68]. In this case, from Equation (6.7) one gets $B_c = 0.2$ T. A more accurate estimate for this example can be obtained by taking into account that non-thermal velocities v_{nonth} in the solar chromosphere can be ~several tens of km/s, so that the total velocity $v_{tot} = (v_T^2 + v_{nonth}^2)^{1/2} - (15–30)$ km/s. Then, $E_L = E_{imax}$ already at B~0.05 Tesla, while B can reach 0.4 T in sunspots.

Another example is edge plasmas of tokamaks. For a low-density discharge in Alcator C-Mod [60], where $N_e \sim 3 \times 10^{13}$ cm^{-3} and $T \sim 5 \times 10^4$ K, we get $B_c = 4$ T, while the actual magnetic field was 8 T.

The ratio of the Zeeman width of hydrogen lines $\Delta\omega_Z$ to the corresponding "halfwidth" of the n-multiplet due to the Lorentz broadening $\Delta\omega_L$.

$$\Delta\omega_Z = eB/(2m_ec), \ \Delta\omega_L = 3n(n-1)\hbar Bv_T/(2m_eec), \ \Delta\omega_Z/\Delta\omega_L = 5680/[n(n-1)T^{1/2}] \qquad (6.10)$$

where n is the principal quantum number of the upper level that is involved in the radiation transition and the atomic temperature T is in Kelvin. For example, for the typical temperature at the edge plasmas of tokamaks $T \sim 5 \times 10^4$ K, Equation (6.10) yields $\Delta\omega_Z/\Delta\omega_L = 25.4/[n(n-1)]$: so that the Lorentz width exceeds the Zeeman width for hydrogen lines of n > 5, while Balmer lines up to n = 16 were observed, e.g., at Alcator C-Mod [60]. Another example: for the typical temperature in the solar chromosphere $T \sim 10^4$ K, Equation (6) yields $\Delta\omega_Z/\Delta\omega_L = 56.8/[n(n-1)]$. So, the Lorentz width exceeds the Zeeman width for hydrogen lines of n > 8, while the Balmer lines up to n~30 were observed [64,68].

Thus, the analytical results from papers [59,66] for the Lorentz–Doppler profiles have sufficiently broad practical applications.

7. Revision of the Inglis-Teller Diagnostic Method

In non-turbulent magnetized plasmas, the Lorentz broadening can predominate over other broadening mechanisms for highly-excited hydrogen lines. In the previous section, it was shown that the Lorentz broadening can significantly exceed both the Stark broadening by the plasma microfield and the Zeeman splitting for high-n hydrogen lines. Below, is the estimate the ratio of the Lorentz and Doppler broadenings from paper [59].

The Doppler Full Width and Half Maximum (FWHM) is

$$(\Delta\omega_D)_{1/2} = 2(\ln2)^{1/2}\omega_0 v_T/c = 1.665\omega_0 v_T/c \qquad (7.1)$$

where ω_0 is the unperturbed frequency of the spectral line. For highly-excited hydrogen lines, where $n_\alpha \gg n_\beta$, one can use the expression $\omega_0 = m_e e^4/(2n_\beta^2\hbar^3)$.

The Lorentz field E_L is confined in the plane perpendicular to **B** where it has the following distribution

$$W_L(E_L)dE_L = (2E_L/E_{LT}^2) \exp(-E_L^2/E_{LT}^2)dE_L, \ E_{LT} = v_TB/c \qquad (7.2)$$

Here, E_{LT} is the average Lorentz field expressed via the thermal velocity v_T of the radiating atoms of mass M. The distribution W_L actually reproduces the shape of the two-dimensional Maxwell distribution of atomic velocities in the plane perpendicular to **B**. This is because the absolute value of the Lorentz field $\mathbf{E}_L = \mathbf{v} \times \mathbf{B}/c$ is $E_L = v_R B/c$, where v_R is the component of the atomic velocity perpendicular to $\mathbf{B}$.

The Lorentz-broadened profile of a Stark component of a hydrogen line reproduces the shape of the Lorentz field distribution from Equation (7.2)

$$S_{\alpha\beta}(\Delta\omega) = (2\Delta\omega/\Delta\omega_{L\alpha\beta}{}^2) \exp(-\Delta\omega^2/\Delta\omega_{L\alpha\beta}{}^2), \Delta\omega_{L\alpha\beta} = kX_{\alpha\beta}Bv_T/c \qquad (7.3)$$

Its FWHM is

$$(\Delta\omega_{L\alpha\beta})_{1/2} = 2.715\,\Delta\omega_{L\alpha\beta} \qquad (7.4)$$

The corresponding FWHM $(\Delta\omega_L)_{1/2}$ of the entire hydrogen line can be estimated by using in Equations (7.3) and (7.4) the average value $< k\,|\,X_{\alpha\beta}\,|> = (n_\alpha{}^2 - n_\beta{}^2)\hbar/(m_e e)$, which for $n_\alpha \gg n_\beta$ becomes $<k\,|\,X_{\alpha\beta}\,| \geq n_\alpha{}^2\hbar/(m_e e)$. Therefore, for the ratio of the FWHM by these two broadening mechanisms one gets

$$(\Delta\omega_L)_{1/2}/(\Delta\omega_D)_{1/2} = n_\alpha{}^2 n_\beta{}^2 B(\text{Tesla})/526 \qquad (7.5)$$

It should be noted that this ratio does not depend on the temperature.

For Balmer lines ($n_\beta = 2$) Equation (7.5) becomes

$$(\Delta\omega_L)_{1/2}/(\Delta\omega_D)_{1/2} = n_\alpha{}^2 B(\text{Tesla})/131 \qquad (7.6)$$

So, e.g., for the egde plasmas of tokamaks, where Balmer lines of $n_\alpha \sim (10\text{--}16)$ have been observed, the Lorentz broadening dominates over the Doppler broadening when the magnetic field exceeds the critical value $B_c \sim 1$ Tesla. This condition is fulfilled in the modern tokamaks and it will be fulfilled also in the future tokamaks.

Another example: in solar chromosphere, where Balmer lines of $n_\alpha \sim (25\text{--}30)$ have been observed, the Lorentz broadening dominates over the Doppler broadening when the magnetic field exceeds the critical value $B_c \sim (0.15\text{--}0.2)$ Tesla. This condition can be fulfilled in sunspots where B can be as high as 0.4 Tesla.

Therefore, it is practically useful to calculate pure Lorentz-broadened profiles of highly-excited Balmer lines. This has been done in paper [59]. By the calculating profiles of an extensive set of hydrogen lines, the following practically important result was derived in [59] for highly excited Balmer lines.

For any two adjacent high-n Balmer lines (such as, e.g., H_{16} and H_{17}, or H_{17} and H_{18}), the sum of their half widths at half maximum in the frequency scale, the sum being denoted here simply as $\Delta\omega_{1/2}$, turned out to be

$$\Delta\omega_{1/2} = A\,[3n^2\hbar Bv_T/(2m_e ec)] \qquad (7.7)$$

where the constant A depends on the direction of observation, as follows:

$$A = 0.80 \text{ (observation perpendicular to } \mathbf{B}) \qquad (7.8)$$

$$A = 1.00 \text{ (observation parallel to } \mathbf{B}) \qquad (7.9)$$

$$A = 0.86 \text{ (“isotropic” observation)} \qquad (7.10)$$

Here, by the "isotropic" observation is meant the situation where along the line of sight there are regions with various directions of the magnetic field, which could be sometimes the case in astrophysics.

The results that are presented in Equations (7.7)–(7.10) lead to a revision of the diagnostic method based on the principal quantum number n_{max} of the last observed line in the spectral series of hydrogen lines, such as, e.g., Lyman, or Balmer, or Paschen lines (though typically Balmer lines are used). This method was first proposed by Inglis and Teller [69]. The idea of the method was that the Stark broadening of hydrogen lines by the ion microfield (in case it is quasistatic) in the spectral series scales as $\sim n^2$. Therefore, at some value $n = n_{max}$, the sum of the Stark half widths at half maximum of the two adjacent lines becomes equal to the unperturbed separation of these two lines, so that they (and the higher lines) merge into a quasicontinuum. Since the Stark broadening is controlled by the ion density N_i (equal to the electron density N_e for hydrogen plasmas), this had led previously to the following simple reasoning.

At the electric field E, for the multiplet of the principal quantum number $n \gg 1$, the separation $\Delta\omega(n)$ of the most shifted Stark sublevel from the unperturbed frequency $\omega_0(n)$ is $\Delta\omega(n) = 3n^2\hbar E/(2m_e e)$. Then, the sum of the "halfwidths" of the two adjacent Stark multiplets of the principal quantum numbers n and $n + 1$ is

$$\Delta\omega_{1/2}(n) = 3n^2\hbar E/(m_e e) \tag{7.11}$$

The unperturbed separation (in the frequency scale) between the hydrogen spectral lines, originating from the highly-excited levels n and $n + 1$ is

$$\omega_0(n + 1) - \omega_0(n) = m_e e^4/(n^3\hbar^3) \tag{7.12}$$

By equating (7.11) and (7.12), one finds

$$n_{max}{}^5 \, E = E_{at}/3 = 5.71 \times 10^6 \text{ CGS}, \ E_{at} = m^2 e^5/\hbar^4 \tag{7.13}$$

($E_{at} = 1.714 \times 10^7$ CGS $= 5.142 \times 10^9$ V/cm is the atomic unit of electric field). For the field E, Inglis and Teller [69] used the most probable field of the Holtsmark distribution, which they estimated as $E_{imax.} = 3.7eN_i{}^{2/3} = 3.7eN_e{}^{2/3}$, and obtained from Equation (7.13) the following relation:

$$N_e n_{max}{}^{15/2} = 0.027/a_0{}^3 = 1.8 \times 10^{23} \text{ cm}^{-3} \tag{7.14}$$

where a_0 is the Bohr radius. It should be noted that Hey [70], by using a more accurate value of the most probable Holtsmark field $E_{imax} = 4.18eN_e{}^{2/3}$, obtained a slightly more accurate numerical constant in the right side of Equation (7.14), namely $0.0225/a_0{}^3$, while Griem [12] suggested this constant to be even twice smaller.

Thus, Inglis-Teller relation (7.14) constituted a simple method for measuring the electron density by the number n_{max} of the observed lines of a hydrogen spectral series. The simplicity of this method is the reason why, despite the existence of more sophisticated (but more demanding experimentally) spectroscopic methods for measuring N_e, this method is still used in both laboratory and astrophysical plasmas. For example, Welch et al. [60] used it (with the constant in the right side of Equation (7.14), as suggested by Griem [12]) for determining the electron density in the low-density discharge at Alcator C-Mod.

However, in magnetized plasmas the Lorentz field E_L can significantly exceed the most probable Holtsmark field E_{max}, as shown in Section 6 of this review (following paper [59]). In this situation, the number n_{max} of the last observable hydrogen line will not be controlled by the electron density, but rather by different parameters, as shown below. Let us first conduct a simplified reasoning along the approach of Inglis and Teller [69]. By substituting $E = E_L$ in the left side of Equation (7.13), the following relation was obtained in paper [59]:

$$n_{max}{}^{10}B^2 \, T(K) = 1.78 \times 10^{18} \, M/M_p \text{ or } n_{max}{}^{10}B^2 \, T(eV) = 1.54 \times 10^{14} \, M/M_p \tag{7.15}$$

where B is the magnetic field in Tesla; M and M_p are the atomic and proton masses, respectively.

More accurate relations were derived in [59] while using the results of the calculations of Lorentz-broadened profiles of high-n Balmer lines by the author of [59] and the corresponding formulas (7.7)–(7.10) for the sum of the half widths at half maximum of two adjacent Balmer lines. In this more accurate, the following relations were obtained in [59].

For the observation perpendicular to **B**:

$$n_{max}{}^{10}B^2 \, T(K) = 2.79 \times 10^{18} \, M/M_p \text{ or } n_{max}{}^{10}B^2 \, T(eV) = 2.40 \times 10^{14} \, M/M_p \qquad (7.16)$$

For the observation parallel to **B**:

$$n_{max}{}^{10}B^2 \, T(K) = 1.78 \times 10^{18} \, M/M_p \text{ or } n_{max}{}^{10}B^2 \, T(eV) = 1.54 \times 10^{14} \, M/M_p \qquad (7.17)$$

For the "isotropic" case (the meaning of which was explained after Equation (7.10)):

$$n_{max}{}^{10}B^2 \, T(K) = 2.38 \times 10^{18} \, M/M_p \text{ or } n_{max}{}^{10}B^2 \, T(eV) = 2.05 \times 10^{14} \, M/M_p \qquad (7.18)$$

Thus, the above formulas, by using the observable quantity n_{max}, allow to measure the atomic temperature T, if the magnetic field is known, or the magnetic field B, if the temperature is known.[2]

8. Stark Broadening of Hydrogen/Deuterium Spectral Lines by a Relativistic Electron Beam: Analytical Results and Applications to Magnetic Fusion

8.1. Preamble

The interaction of a Relativistic Electron Beam (REB) with plasmas has both the fundamental importance for understanding physics of plasmas and practical applications. The latter include (but not limited to) plasma heating, inertial fusion, generation of high-intensity coherent microwave radiation, acceleration of charged particles in plasmas—see, e.g., papers [73–75], and references therein.

The latest (though negative) application relates to magnetic fusion and deals with runaway electrons. In some discharges in tokamaks, the plasma current decays and it is partly replaced by runaway electrons that reach relativistic energies: this poses danger to the mission of the next generation tokamak ITER—see, e.g., papers [76–78] and references therein. At various discharges at different tokamaks, such as, e.g., those presented in papers [79–81], the energy of the runaway electrons was measured in the range ~(0.2–10) MeV and the ratio of their density to the density of the bulk plasma electrons was measured in the range ~$(10^{-1}–10^{-4})$.

Therefore developing diagnostics of a REB and its interaction with plasmas should be important. In the particular case of tokamaks, the development of a REB should be timely detected to allow the mitigation of the problem.

Diagnostics based on the analysis of spectral line shapes have known advantages over others. They are not intrusive and allow measuring plasma parameters and parameters of various fields in plasmas without perturbing the parameters to be measured—see, e.g., books [1–7].

In paper [82], there was presented a theory of the Stark broadening of hydrogen/deuterium spectral lines by a REB. The theory was developed analytically by using an advanced formalism. The authors of paper [82] discussed the possible application of these analytical results to magnetic fusion edge plasmas, taking into account also the major outcome of the interaction of a REB with plasmas: the development of strong Langmuir waves.[3]

---

[2] A shorter version of the present paper was published in 2013 in the fast track journal IRAMP [71], a publication in which does not prevent publishing a more extended version elsewhere. We note that later, in 2014, Rosato et al. published a paper [72], where they reinvented some results from our paper [71] concerning the principal quantum number n_{max} of the last observable hydrogen spectral line without referring to paper [71]. Namely, they reinvented our preliminary approximate formula for the product $n_{max}{}^{10}B^2 \, T$ (where B is the magnetic field and T is the atomic temperature), but did not come up with our more accurate results for $n_{max}{}^{10}B^2 \, T$ based on our calculations of Lorentz-broadened profiles of high-n Balmer lines. Just as in our paper [71], Rosato et al. applied their results to magnetic fusion, but they did not apply their results to solar physics—in distinction to our paper [71].

[3] We note that Rosato et al. [83] attempted studying the Stark broadening of hydrogen line by a REB in magnetic fusion edge plasmas. However, they used the quasistatic approximation, which is totally inappropriate for the broadening by fast electrons of a REB (it is inappropriate even for the broadening by thermal electrons in such plasmas).

8.2. Analytical Results and Applications to Magnetic Fusion

The presence of a REB introduces anisotropy in the process of the Stark broadening of spectral lines in plasmas. A different kind of anisotropic Stark broadening was first considered by Seidel in 1979 [21] for the following situation. If hydrogen atoms radiate from a plasma consisting mostly of much heavier ions, then in the reference frame moving with the velocity v of the radiating hydrogen atom, the latter "perceives" a beam of the much heavier ions moving with the velocity v. Seidel [21] treated this situation by applying the so-called standard (or conventional) theory of the impact broadening of hydrogen lines, which is also known as Griem's theory [13]. Therefore, while Seidel [21] should be given credit for pioneering the anisotropic Stark broadening, his specific calculations had a weakness that plagues the standard theory: the inherent divergence at small impact parameters causing the need for a cutoff defined only by an order of magnitude.

Later in paper [24], the authors considered the same situation as Seidel [21], but applied a more advanced theory of the Stark broadening, called the generalized theory that is developed in paper [84] and also presented = in books [5,7]. (It should be emphasized that, in paper [24], it was the application of the "core" generalized theory from paper [84] without the additional effects that were introduced later and were the subject of discussions in the literature.) The authors of paper [24] took into the exact account (in all the orders of the Dyson expansion) the projection of the dynamic, heavy-ion-produced electric field onto the velocity of the radiator exactly. As a result, there was no divergence at small impact parameters, and thus no need for the imprecise cutoff.

In paper [82], the results from which we outline below, the authors used the formalism from paper [24] to treat the Stark broadening of hydrogen/deuterium spectral lines by a REB in plasmas. There are two major distinctions from paper [24]: (1) the broadening is by a beam of electrons rather than ions; (2) the electrons are relativistic.

The relativistic counterparts C_{r+} and C_{r-} of the broadening functions C_+ and C_-, as calculated in [82] by the core generalized theory from [84], became as follows:

$$C_{r\pm}(Z) = \frac{1}{2\gamma^4} \int\limits_{-\infty}^{\infty} \int\limits_{-\infty}^{x_1} \frac{dx_1 dx_2}{[g_r(x_1)g_r(x_2)]^3} \exp\left[\frac{i}{Z}(1/g_r(x_1) \pm 1/g_r(x_2))\right],$$
$$g_r(x) \equiv \sqrt{1/\gamma^2 + x^2}. \tag{8.1}$$

Here, Z is the scaled impact parameter:

$$Z = 2m_e v\rho/(3n\hbar) \tag{8.2}$$

where n is the principal quantum number of the upper level and ρ is the impact parameter; the quantity

$$\gamma = 1/(1 - v^2/c^2)^{1/2} \tag{8.3}$$

is the relativistic factor. For the real parts $A_{r\pm} = \mathrm{Re}\, C_{r\pm}$, the double integral in Equation (8.1) can be calculated analytically. It yields:

$$A_{r-} = (\pi/2)^2 \left[\mathbf{H}_{-1}(1/s) + J_1(1/s)\right],\ A_{r+} = (\pi/2)^2 \left[\mathbf{H}_{-1}(1/s) - J_1(1/s)\right],\ s = Z/\gamma \tag{8.4}$$

where $\mathbf{H}_{-1}(1/s)$ and $J_1(1/s)$ are Struve and Bessel functions, respectively. Below, we omit the suffix "r" for brevity.

The width of spectral line components is controlled by the subsequent integral over the scaled impact parameter Z:

$$a_\pm = \int\limits_0^{Z_{max}} A_\pm(Z)dZ/Z = \int\limits_0^{Z_{max}/\gamma} A_\pm(s)ds/s,\ s = Z/\gamma \tag{8.5}$$

24

Figure 9 shows the plot of the integrand $A_-(s)/s$ versus s. It is seen that the corresponding integral a_- does not diverge at small impact parameters.

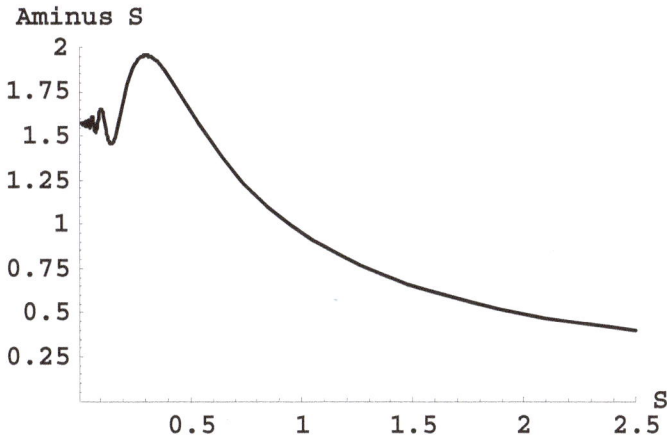

Figure 9. The integrand $A_-(s)/s$, corresponding to the widths function a_-, versus $s = Z/\gamma$, where Z is the scaled impact parameter defined by Equation (8.2) and γ is the relativistic factor defined by Equation (8.3) [82].

Figure 10 presents the plot of the integrand $A_-(s)/s$ versus s and Figure 11 shows a magnified part of this plot at small impact parameters. It is seen that the corresponding integral a_+ also does not diverge at small impact parameters.

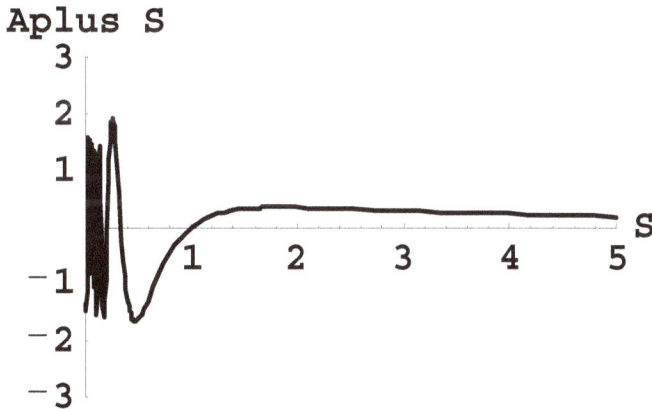

Figure 10. The integrand $A_+(s)/s$, corresponding to the widths function a_+, versus $s = Z/\gamma$, where Z is the scaled impact parameter defined by Equation (8.2) and γ is the relativistic factor, as defined by Equation (8.3) [82].

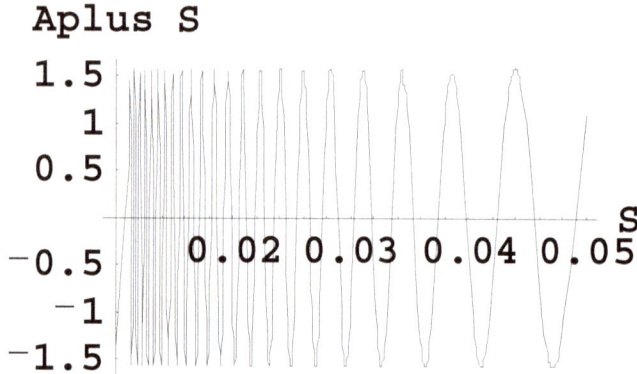

Figure 11. Same as in Figure 10, but for small impact parameters [82].

Thus, the integrals over the scale impact parameter Z in Equation (8.5) converge at small impact parameters—in distinction to what would have resulted from the standard theory. At large Z, the integral diverge (just as what would have resulted from the standard theory), which is physically because of the long-range nature of the Coulomb interaction between the charged particles. However, due to the Debye screening in plasmas, there is a natural upper cutoff Z_{max}:

$$Z_{max} = uZ_0, \ u = v/c = (1 - 1/\gamma^2)^{1/2}, \ Z_0 = 2m_ec\rho_D/(3n\hbar) \tag{8.6}$$

Here

$$\rho_D = [T_e/(4\pi e^2 N_e)]^{1/2} \tag{8.7}$$

is the Debye radius; T_e and N_e are the temperature and the density of bulk electrons, respectively.

The integration in Equation (8.5) can be performed analytically because the integrals in Equation (8.5) have the following antiderivatives

$$j_\pm(s) = \int A_\pm(s)ds/s = (\pi^2/8)\{ (2/\pi)\text{MeijerG}[\{\{0\},\{1\}\},\{\{0,0\},\{-1/2,1/2\}\},1/(4s^2] + \mathbf{H}_{-1}{}^2(1/s) $$
$$+ \mathbf{H}_0{}^2(1/s) \pm [1 - {}_1F_2(1/2;1,2;-1/s^2)] \tag{8.8}$$

where MeijerG[...] and ${}_1F_2(\ ... \ )$ are the MeijerG function and the generalized hypergeometric function, respectively. Thus, the following analytical results for the width functions were obtained in [82]:

$$a_\pm = j_\pm(Z_{max}/\gamma) - j_\pm(0) \tag{8.9}$$

As an example, the authors of paper [82] explicitly calculated the shape $I(\Delta\omega,\gamma)$ of the spectral line Ly-alpha broadened by a REB, where $\Delta\omega$ is the detuning from the unperturbed frequency of the spectral line. Similarly, to paper [24], after inverting the spectral operator, they obtained:

$$I(\Delta\omega,\gamma) = \frac{1}{3\pi}\left(\frac{\Gamma_\pi}{\Delta\omega^2 + \Gamma_\pi^2} + \frac{2\Gamma_\sigma}{\Delta\omega^2 + \Gamma_\sigma^2}\right), \tag{8.10}$$

where Γ_π and Γ_σ are the half-widths at half-maximum of the π- and σ-components of the Ly-alpha line, respectively. They are expressed, as follows:

$$\Gamma_\sigma = [\eta_0/(1 - 1/\gamma^2)^{1/2}][\ j_-(Z_{max}/\gamma) - j_-(0)] \tag{8.11}$$

$$\Gamma_\pi = \left[\eta_0/\left(1 - 1/\gamma^2\right)^{1/2}\right] \int_0^\infty [A_-(s) - A_+(s)]ds/s \tag{8.12}$$

where

$$\eta_0 = 4\pi\hbar^2 N_e/(m_e^2 c) = 5.618 \times 10^{-10} \, N_e(\text{cm}^{-3}) \, \text{s}^{-1} \qquad (8.13)$$

It is worth noting that, in Equation (8.12), the upper limit of the integration is infinity. This is because for the π-component of the Ly-alpha line the width in Equation (8.12) is proportional to the difference of diagonal and nondiagonal matrix elements of the broadening operator, so that the corresponding integral converges not only at small, but also at large impact parameters, yielding the following relatively simple expression for the width:

$$\Gamma_\pi = \pi^2\eta_0/[4(1 - 1/\gamma^2)^{1/2}] \qquad (8.14)$$

Figure 12 (which reproduces Figure 4 from [82]) shows the plot of the scaled width of the σ-component Γ_σ/η_0 (upper curve) and of the scaled width of the π-component Γ_π/η_0 (lower curve) of the Ly-alpha line that is broadened by a REB versus the relativistic factor γ at $N_e = 10^{15}$ cm^{-3} and $T_e = 2$ eV. It is seen that as γ increases from unity, both widths significantly decrease.

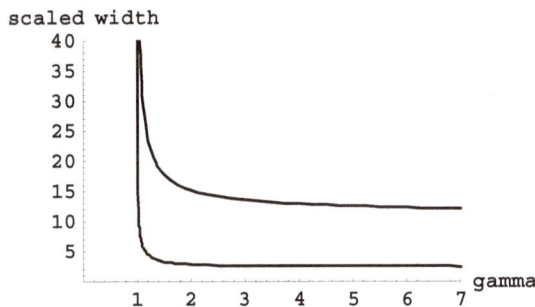

Figure 12. The scaled width of the σ-component Γ_σ/η_0 (upper curve) and the scaled width of the π-component Γ_π/η_0 (lower curve) of the Ly-alpha line broadened by a Relativistic Electron Beam (REB) versus the relativistic factor γ at $N_e = 10^{15}$ cm^{-3} and $T_e = 2$ eV [82].

Figure 13 (which reproduces Figure 5 from [82]) presents the ratio Γ_σ/Γ_π versus the relativistic factor γ at $N_e = 10^{15}$ cm^{-3} and $T_e = 2$ eV. It is seen that as γ increases from unity, this ratio increases, then reaches the maximum, and then decreases. The maximum ratio $\Gamma_\sigma/\Gamma_\pi = 5.39$ corresponds to $\gamma = 2^{1/2}$.

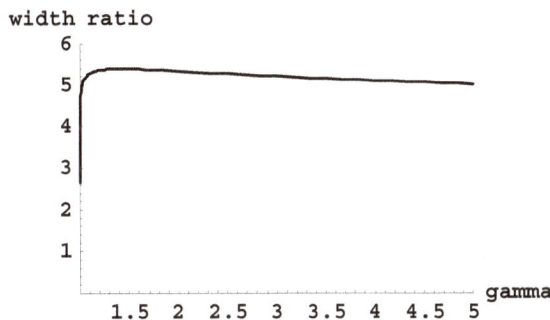

Figure 13. Ratio Γ_σ/Γ_π of the widths of the σ- and π-components of the Ly-alpha line versus the relativistic factor γ at $N_e = 10^{15}$ cm^{-3} and $T_e = 2$ eV [82].

Separate measurements of the widths of the σ- and π-components (and thus of the ratio Γ_σ/Γ_π) can be performed for the observation perpendicular to the REB velocity by placing a polarizer into the optical system: when the axis of the polarizer would be perpendicular or parallel to the REB velocity, then one would be able to measure Γ_σ or Γ_π, respectively. By monitoring the dynamics of the ratio Γ_σ/Γ_π, it would be possible, at least in principle, to detect the development of a REB in tokamaks and to engage the mitigation of the problem.

Figure 14 (which reproduces Figure 6 from [82]) shows the theoretical profiles of the entire Ly-alpha line, corresponding to the observation that is perpendicular to the REB velocity without the polarizer, at $N_e = 10^{15}$ cm^{-3} and $T_e = 2$ eV. The profiles were calculated using Equations (8.10)–(8.13) and presented versus the scaled detuning $\Delta\omega/\Gamma_\pi$ denoted as d. Due to the scaled detuning, the profiles are "universal" in the sense that they are independent of the beam electron density. The solid curve corresponds to $\gamma = 2^{1/2}$, while the dashed curve—to $\gamma = 1$. In the case of $\gamma = 2^{1/2}$, the profile is by 12% narrower than for the case of $\gamma = 1$. Detecting the development of a REB via such relatively small decrease of the width seems to be less advantageous when compared to the polarization analysis of the width that is discussed above, where the widths ratio Γ_σ/Γ_π could increase by an order of magnitude as a REB develops in the plasma.

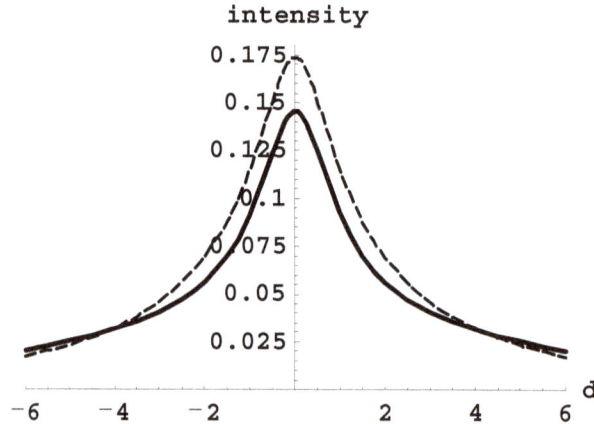

Figure 14. Theoretical profiles of the entire Ly-alpha line, corresponding to the observation perpendicular to the REB velocity without the polarizer, at $N_e = 10^{15}$ cm^{-3} and $T_e = 2$ eV. The profiles were calculated using Equations (8.10)–(8.13) and presented versus the scaled detuning $\Delta\omega/\Gamma_\pi$ denoted as d. The solid curve corresponds to $\gamma = 2^{1/2}$, while the dashed curve – to $\gamma = 1$ [82].

The above theoretical results represented the Stark broadening of hydrogen/deuterium spectral lines only by a REB without allowing for other factors affecting the lineshapes. This was done in the first part of paper [82] for presenting the effect of a REB on the lineshape in the "purest" form. In the second part of [82], the authors removed this restriction, as follows[4].

---

[4] So far, we used, as an example the Ly-alpha line just to get the message across (since we obtained relatively simple analytical expressions for the shape of this line). We note that at $N_e \sim 10^{15}$ cm^{-3}, the Stark width of the Lyman-alpha line calculated by Equations (8.11)–(8.14) would be by about two and a half orders of magnitude below the natural width. However, the dynamical Stark width scales $\sim n^4$, while the natural width scales $\sim 1/n^5$ (n being the principal quantum number). Therefore, for the lines that are originating from the level of n = 4 (such as Ly-gamma, Balmer-beta, Paschen-alpha) and higher levels, the corresponding dynamical Stark width would exceed the natural width.

The major outcome of the interaction of a REB with plasmas is the development of strong Langmuir waves—see, e.g., book [85]. The maximum amplitude E_0 of the Langmuir wave electric field is [85]:

$$E_0 = [8\pi m_e c^2 \gamma^2 N_b^{4/3} / N_e^{1/3}]^{1/2}, \gamma N_b^{1/3} / N_e^{1/3} \ll 1 \qquad (8.15)$$

For the case of $N_e = 10^{15}$ cm^{-3}, $N_b = 6 \times 10^9$ cm^{-3}, and $\gamma = 2^{1/2}$, corresponding to an early stage of the development of a REB in tokamaks, Equation (8.14) yields $E_0 = 20$ kV/cm.

The primary manifestation of Langmuir waves in the profiles of hydrogen/deuterium or hydrogenlike spectral lines is the appearance of some local structures (called L-dips) at certain locations of the spectral line profile. This phenomenon arises when radiating atoms/ions are subjected simultaneously to a quasistatic field **F** and to a quasimonochromatic electric field **E**(t) at the characteristic frequency ω, where E < F. In the heart of this phenomenon is the dynamic resonance between the Stark splitting of hydrogenic spectral lines and the frequency ω or its harmonics. There is a rich physics behind the L-dip phenomenon: even when the applied electric field is monochromatic, there occurs a nonlinear dynamic resonance of multifrequency nature involving all of the harmonics of the applied field—as it was explained in detail in paper [86]. Further details on the theory of the L-dips can be found in books [1,7].

As for the experimental studies of the L-dips, books [1,7] and later reviews [87–89] summarize all such studies with applications to plasma diagnostics. The practical significance of studies of the L-dips is threefold. First, they provide the most accurate passive spectroscopic method for measuring the electron density N_e in plasmas, e.g., more accurate than the measurement from the line broadening. This passive spectroscopic method for measuring N_e does not differ in its high accuracy from the active spectroscopic method—more complicated experimentally—using the Thompson scattering [90]. Second, they provide the only one non-perturbative method for measuring the amplitude of Langmuir waves in plasmas [1]. Third, in laser-produced plasmas, they facilitate revealing physics behind the laser-plasma interaction [91–93].

The resonance between the Stark splitting of hydrogenic spectral lines and the frequency ω of the Langmuir wave (which is close to the plasma electron frequency ω_{pe}) or its harmonics, translates into specific locations of L-dips in spectral line profiles, which depend on N_e since ω_{pe} depends on N_e. In particular, for relatively low density plasmas (like in magnetic fusion machines) or in the situation, where the quasistatic field **F** is dominated by the low-frequency electrostatic turbulence (e.g., the ion acoustic turbulence), for the Ly-lines, the distance of an L-dip from the unperturbed wavelength λ_0 can be expressed as

$$\Delta\lambda_{dip}(qk, N_e) = [\lambda_0^2 / (2\pi c)] qk\omega_{pe} (N_e) \qquad (8.16)$$

Here, λ_0 is the unperturbed wavelength of the spectral line and $q = n_1 - n_2$ is the electric quantum number that is expressed via the parabolic quantum numbers n_1 and n_2: $q = 0, \pm1, \pm2, \dots, \pm(n-1)$. The electric quantum number labels Stark components of Ly-lines. Equation (8.16) shows that for a given electron density N_e, the locations of L-dips are controlled by the product qk.

It should be emphasized that the abbreviation "L-dip" refers to a local structure consisting of the central minimum and (generally) two adjacent bumps surrounding the central minimum—the latter is called "dip" for brevity. Equation (8.16) specifies the locations of the central minima (dips) of these structures: it is from the locations of the central minima that the electron density can be determined experimentally. The dip-bump separation is controlled by the Langmuir field amplitude E_0 and thus allows for the experimental determination of E_0 [1].

For finishing this brief excerpt from the L-dip theory necessary for understanding the next paragraphs, it should be also noted that when a bump-dip-bump structure is superimposed with the *inclined* part of the spectral line profile, this might lead to the appearance of a secondary minimum of no physical significance. Also, when the L-dip is too close to the unperturbed wavelength, its bump that is nearest to the unperturbed wavelength might have zero or little visibility. These subtleties were observed numerous times [1,87–89] and will also be relevant below.

So, the authors of [82] used the Ly-delta line of deuterium as an illustrative example of possible diagnostics of the early stage of the development of a REB in tokamaks. The Ly-delta line has four Stark components in each wing, corresponding to q = ±1, ±2, ±3, ±4. Therefore, according to Equation (8.16), the L-dip in the profile of the component of q = 1 due to the four-quantum resonance (k = 4) coincides by its location with the L-dip in the profile of the component of q = 2 due to the two-quantum resonance (k = 2), and with the L-dip in the profile of the component of q = 4 due to the one-quantum resonance (k = 1). The superposition of three different L-dips at the same location results in the L-super-dip of the significantly enhanced visibility.

Also, according to Equation (8.16), the L-dip in the profile of the component of q = 1 due to the two-quantum resonance (k = 2) coincides by its location with the L-dip in the profile of the component of q = 2 due to the one-quantum resonance (k = 1). The superposition of two different L-dips at the same location results also enhances the visibility of the resulting structure.

For diagnostic purposes, it is important to choose the spectral line where superpositions of several L-dips at the same location in the profile are expected. This is because due to competing broadening mechanisms (such as, e.g., the dynamical broadening by electrons and some ions, as well as the Doppler broadening), a single L-dip could be washed out, but a superposition of two or especially three L-dips at the same location could "survive" the competition.

Figure 15 (which reproduces Figure 7 from [82]) presents the theoretical profile of the Ly-delta line of deuterium, calculated with the allowance for all broadening mechanisms and for the effect of strong Langmuir waves (in distinction to the Ly-alpha profile in Figure 14 that illustrated the pure effect of the broadening by the REB only), at the following parameters: $N_e = 10^{15}$ cm^{-3}, $N_b = 6 \times 10^9$ cm^{-3}, $\gamma = 2^{1/2}$ (corresponding to the beam kinetic energy of 210 keV), and $T_e = 2$ eV. The solid curve corresponds to the presence of the strong Langmuir waves of $E_0 = 20$ kV/cm that are caused by a REB (according to Equation (8.16)), while the dashed curve corresponds to the absence of the REB. The detuning $\Delta\lambda$ (denoted "dlambda" in Figure 15) is in Angstrom.

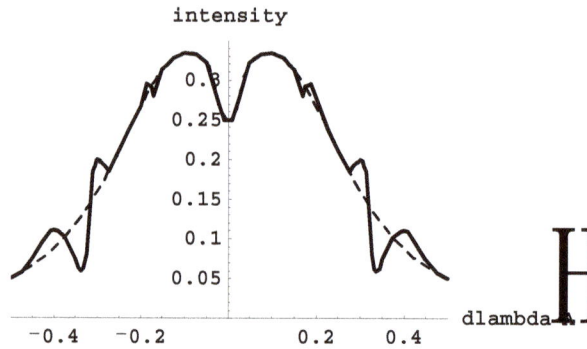

Figure 15. Theoretical profile of the Ly-delta line of deuterium, calculated with the allowance for all broadening mechanisms and for the effect of strong Langmuir waves, at the following parameters: $N_e = 10^{15}$ cm^{-3}, $N_b = 6 \times 10^9$ cm^{-3}, $\gamma = 2^{1/2}$ (corresponding to the beam kinetic energy of 210 keV), and $T_e = 2$ eV. The solid curve corresponds to the presence of the strong Langmuir waves of $E_0 = 20$ kV/cm caused by a REB, while the dashed curve corresponds to the absence of the REB. The detuning $\Delta\lambda$ (denoted "dlambda") is in Angstrom [82].

The theoretical profile shown by the solid curve exhibits two L-dip structures at both the red and blue parts of the profile. The central minimum of the L-super-dips of qk = ±4 is at $\Delta\lambda = \pm0.338$ A. This L-super-dip structure is very pronounced: the central minimum is relatively deep and both of the adjacent bumps are clearly visible. (Being superimposed with the inclined part of the profile, it creates also secondary minima of no physical significance at $\Delta\lambda = \pm0.275$ A).

The L-dip structure of $qk = \pm2$, whose central minimum is at $\Delta\lambda = \pm0.169$ A, is also visible. However, it is less pronounced (as compared with the L-super-dip of $qk = \pm4$) and its bump closest to the unperturbed wavelength has practically zero visibility. This is due to the fact that because of the proximity of this L-dip to the unperturbed wavelength, the ion dynamical broadening is more significant than for the L-super dip at $\Delta\lambda = \pm0.338$ A.

In this example, the ratio of the energy density of the Langmuir waves to the thermal energy density of the plasma (the ratio called sometimes the "degree of the turbulence") was $E_0^2/(8\pi N_e T_e)\sim0.06$. Since $E_0^2/(8\pi N_e T_e) >> m_e/M\sim0.0003$ (where M is the mass of deuterium atoms), these Langmuir waves qualify as the strong turbulence.

Thus, the monitoring the shape of deuterium spectral lines (such as, e.g., Ly-delta, or Balmer-beta, or Balmer-delta, or Paschen-beta, or Paschen-delta) and the observation of the formation of the L-dips in the experimental profile would constitute the detection of the early stage of the development of a REB in tokamaks. The detection of the early stage of the development of a REB would allow for mitigating the problem in a timely manner.

As for the final stage of the development of the REB (where the beam electron density N_b could become just of one or two orders of magnitude below the electron density N_e of bulk electrons), would be manifested—and thus could be detected, at least in principle—by a decrease of the width of hydrogen/deuterium spectral lines. In paper [82], there was demonstrated that especially sensitive to the final stage of the development of the REB would be the ratio of widths of σ- and π-components, which could be determined by the polarization analysis.

9. Influence of Magnetic-Field-Caused Modifications of Trajectories of Plasma Electrons on Shifts and Relative Intensities of Spectral Line Components: Applications to Magnetic Fusion and White Dwarfs

In many types of plasmas—such as, e.g, tokamak plasmas (see, e.g., [60]), laser-produced plasmas (see, e.g., [94]), capacitor-produced plasmas [95], and astrophysical plasmas [96,97]—there are strong magnetic fields. In a strong magnetic field **B**, perturbing electrons basically spiral along magnetic field lines. Therefore, their trajectories are not rectilinear in the case of neutral radiators or not hyperbolic in the case of charged radiators. This was taken into account in paper [98], which we briefly follow here. In paper [98], the presentation started from the general framework for calculating shapes of hydrogen (or deuterium) spectral lines in strongly-magnetized plasmas allowing for spiraling trajectories of perturbing electrons.

In a plasma containing a strong magnetic field **B**, the author of [98] chose the z-axis along **B**. For the case of neutral radiators, the radius-vector of a perturbing electron can be represented in the form

$$\mathbf{R}(t) = \rho\mathbf{e}_x + v_z t\mathbf{e}_z + r_{Bp}\left[\mathbf{e}_x\cos(\omega_B t+\varphi) + \mathbf{e}_y\sin(\omega_B t+\varphi)\right] \tag{9.1}$$

where the x-axis is chosen along the impact parameter vector ρ. Here, v_z is the electron velocity along the magnetic field and

$$r_{Bp} = v_p/\omega_B,\ \omega_B = eB/(m_e c) \tag{9.2}$$

where v_p is the electron velocity in the plane perpendicular to **B**; ω_B is the Larmor frequency.

For the atomic electron to experience the spiraling nature of the trajectories of perturbing electrons, it requires $\rho_{De}/r_{Bp} > 1$, where ρ_{De} is the electron Debye radius. Taking into account that the average over the 2D-Maxwell distribution $<1/v_p> = \pi^{1/2}/v_{Te}$, where $v_{Te} = (2T_e/m_e)^{1/2}$ is the mean thermal velocity of perturbing electrons, the condition $\rho_{De}/r_{Bp} > 1$ can be rewritten in the form $\pi^{1/2}\omega_B/\omega_{pe} > 1$ (where ω_{pe} is the plasma electron frequency) or

$$B > B_{cr} = c(4m_e N_e)^{1/2},\ B_{cr}(\text{Tesla}) = 1.81 \times 10^{-7}\ [N_e(\text{cm}^{-3})]^{1/2} \tag{9.3}$$

where N_e is the electron density. For example, for the edge plasmas in tokamaks, at $N_e = 10^{14}$ cm^{-3}, the condition (9.3) becomes $B > 1.8$ Tesla, which is fulfilled in modern tokamaks. Another example is

DA white dwarfs: at $N_e = 10^{17}$ cm^{-3}, condition (9.3) becomes B > 50 Tesla, which is fulfilled in many DA white dwarfs [99]. Also, the condition (9.3) is easily fulfilled in capacitor-produced plasmas [95] and in plasmas that are produced by high-intensity lasers [94] (in the latter case, radiators would be charged rather than neutral).

We consider the situation where the temperature T of the radiators satisfies the condition

$$T \ll (11.12 \text{ keV}/n)(M/M_H)^2 \tag{9.4}$$

where M is the radiator mass, M_H is the mass of hydrogen atoms, n is the principal quantum number of the energy levels, from which the spectral line originates. Under this condition, the Lorentz field effects can be disregarded compared to the "pure" magnetic field effects.

To get the message across in a relatively simple form, the author of [98] limited himself by the Lyman lines. Then, matrix elements of the electron broadening operator in the impact approximation have the form

$$\Phi_{\alpha\alpha} = N_e \int_{0\to}^{\infty\to} dV_p W_p(V_p) \int_{-\infty+}^{\infty+} dV_Z W_Z(V_Z) V_Z \int_0^{\rho_{max}} d\rho \, 2\pi\rho \, \langle S_{\alpha\alpha} - 1 \rangle, \tag{9.5}$$

where $W_p(v_p)$ is the two-dimensional (2D)-Maxwell distribution of the perpendicular velocities, $W_Z(v_Z)$ is the one-dimensional (1D)-Maxwell (Boltzmann) distribution of the longitudinal velocities, $\rho_{max} = \rho_{De}$ is the maximum impact parameter, S is the scattering matrix, <...> stands for averaging; α and α' label sublevels of the upper energy level involved in the radiative transition.

In the first order of the Dyson expansion one gets

$$S_{\alpha\alpha} - 1 = -i(e^2/\hbar) \int_{-\infty}^{\infty} dt \{ r_{\alpha\alpha} R(t)/[R(t)]^3 \} \exp(i\omega_{\alpha\alpha}t), \tag{9.6}$$

where r is the radius-vector of the atomic electron, $r_{\alpha\alpha'} R$ is the scalar product (also known as the dot-product). Here, $\omega_{\alpha\alpha'}$ is the energy difference between the energy sublevels α and α' divided by $\hbar$; for the adjacent energy sublevels, we have

$$|\omega_{\alpha\alpha'}| = \omega_B/2 \tag{9.7}$$

The following notations were introduced in [98]:

$$s = \rho/r_{Bp}, g = v_Z/v_p, w = v_Z t/r_{Bp} \tag{9.8}$$

In these notations, Equation (9.6) can be rewritten as

$$S_{\alpha\alpha} - 1 = -i[e^2/(\hbar r_{Bp} V_Z)] \int_{-\infty}^{\infty} dw \, \{Z_{\alpha\alpha} w \delta_{\alpha\alpha} + \exp\left[\pm \frac{iw}{2g}\right] [y_{\alpha\alpha} \sin\left(\frac{w}{g} + \varphi\right)$$
$$+ X_{\alpha\alpha}\left(\cos(\frac{w}{g} + \varphi) + s\right)]\}/[1 + w^2 + 2s \, \cos\left(\frac{w}{g} + \varphi\right) + S^2]^{3/2}, \tag{9.9}$$

where $\delta_{\alpha\alpha'}$ is the Kronecker-delta (the parabolic quantization is used here).

In the integrand in Equation (9.9), the terms containing $z_{\alpha\alpha'}$ is the odd function of w, so that the corresponding integral vanishes. The term containing $y_{\alpha\alpha'}$ vanishes after averaging over the phase φ. As for the term containing $x_{\alpha\alpha'}$, after averaging over the phase φ, it becomes as follows (it should be noted that in this setup the angular averaging of vector ρ is irrelevant—in distinction to the rectilinear trajectories).

$$\langle S_{\alpha\alpha} - 1 \rangle = -i\left[e^2/\left(hr_{Bp}V_Z\right)\right] \int_{-\infty}^{\infty} dw \cos\left[\frac{w}{2g}\right] \left\{ K\left[\frac{4s}{j(w,s)}\right] - \frac{E\left[\frac{4s}{j(w,s)}\right]\left(w^2+1-s^2\right)}{\left[w^2+(s-1)^2\right]} \right\} / \left\{ \pi s[j(w,s)]^{\frac{1}{2}} \right\}, \tag{9.10}$$

where $K(u)$ and $E(u)$ are the elliptic integrals, and

$$j(w,s) = w^2 + (1+s)^2 \tag{9.11}$$

In paper [98], it was denoted:

$$u = \rho_{De}/r_{Bp} \tag{9.12}$$

Then, combining Equations (9.5) and (9.10), the first order of the *nondiagonal* elements of the electron broadening operator $\Phi_{\alpha\alpha'}{}^{(1)}$ can be represented in the form (the diagonal elements of $\Phi^{(1)}$ vanish because the diagonal elements of the x-coordinate are zeros):

$$\Phi_{\alpha\alpha}{}^{(1)} = -iX_{\alpha\alpha}\left(e^2/h\right)\rho_{De}N_e \int_0^{\infty} dV_p W_p(V_p) \int_{-\infty}^{\infty} dV_Z W_Z(V_Z) \, f_0\,(g,u), \tag{9.13}$$

where

$$f_0(g,u) = (1/u) \int_0^u ds \int_{-\infty}^{\infty} dw \cos\left[\frac{w}{2g}\right] \left\{ K\left[\frac{4s}{j(w,s)}\right] - \frac{E\left[\frac{4s}{j(w,s)}\right]\left(w^2+1-s^2\right)}{\left[w^2+(s-1)^2\right]} \right\} / \pi s[j(w,s)]^{\frac{1}{2}} \tag{9.14}$$

It should be noted that $\Phi_{\alpha\alpha'}{}^{(1)}$ scales with the electron density as $N_e^{1/2}$.

For obtaining results in the simplest form, in [98] there was calculated the average $\langle g \rangle = \langle v_z/v_p \rangle$

$$\langle g \rangle = \int_{0\rightarrow}^{\infty\rightarrow} dV_p W_p(V_p) \int_{-\infty}^{\infty} dV_Z W_Z(V_Z) V_Z/V_P = 1 \tag{9.15}$$

and denoted $f_0(1, u) = f(u)$. The argument can be represented as $u = B/B_{cr}$, where B_{cr} is the critical value of the magnetic field defined in Equation (9.3). Figure 16 shows this universal function $f(u)$ that controls the phenomenon under consideration.

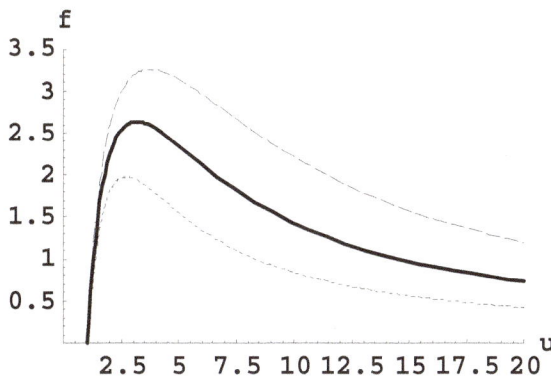

Figure 16. The universal function $f(u)$, controlling the first order of the electron broadening operator, shown at three different values of the ratio $g = v_z/v_p$: $g = 1$ (solid curve), $g = 3/2$ (dashed curve), $g = 2/3$ (dotted curve). The argument $u = B/B_{cr}$, where B_{cr} is the critical value of the magnetic field defined in Equation (9.3) [98].

It is seen that f(u) starts growing at $B/B_{cr} > 1$, reaches the maximum at $B/B_{cr} = 3$, and it then gradually diminishes as B/B_{cr} further increases.

The presence of the maximum of f(u) can be physically understood, as follows. At $\rho_{De} < r_{Bp}$ the effect is absent because the radius of the spirals of the trajectory of perturbing electrons exceeds the Debye radius. In the opposite limit, i.e., at $\rho_{De} >> r_{Bp}$, the radius of the spirals of the trajectory is so small (compared to the Debye radius), that the atomic electron perceives the trajectory almost as a straight line (the straight line of a "width"~r_{Bp}), so that $\Phi_{\alpha\alpha'}^{(1)}$ gradually goes to the limit of zero, i.e., to the limit corresponding to rectilinear trajectories of perturbing electrons.

The real and imaginary parts of $\Phi_{\alpha\alpha'}$ relate to the width and shift of spectral components of hydrogen/deuterium lines, respectively, as it is well-known from the fundamentals of the line broadening theory (see, e.g., books [5,13]). Since the above first order contribution $\Phi_{\alpha\alpha'}^{(1)}$ is purely imaginary, it is clear in advance that it should translate primarily into an *additional shift* of Zeeman components of hydrogen/deuterium lines. While the above analytical results were obtained for Lyman lines, it is obvious that for *any* hydrogen/deuterium line of *any* spectral series (Balmer, Paschen etc.) there should be an additional shift of the Zeeman components of the line that is caused by the spiraling trajectories of perturbing electrons.

In paper [98], the Ly-alpha line was chosen for illustrative examples. It was shown that the primary effect of spiraling trajectories of perturbing electrons is a significant change of the ratio of the intensity of the central peak to the intensity of either one of the two lateral peaks. The secondary effect is an additional shift of the lateral components. Here are some details.

Figure 17 shows the comparison of the spectra of the entire Ly_α line for the example of tokamak T-10 with the allowance for the spiraling trajectories of the perturbing electrons (solid line) and without this allowance (dotted line). The plasma parameters corresponded to the edge plasmas of tokamak T-10 in the experiment that was described in [100]: B = 1.65 Tesla, $N_e = 2 \times 10^{13}$ cm^{-3}, T = (10–15) eV. The spectra were calculated for the observation at the angle of 45 degrees with respect to the magnetic field. The primary effect of the allowance for the spiraling trajectories of the perturbing electrons is that the ratio of the intensity of the central peak to the intensity of any of the two lateral peaks increased by 82% (from 0.56 to 1.02). The secondary effect is that the shift of the lateral (σ-) components increased by 23%.

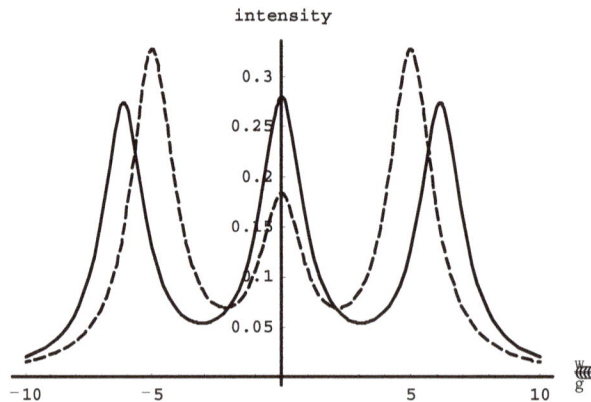

Figure 17. Comparison of the theoretical spectra of the Ly_α line for the above example of tokamak T-10 with the allowance for the spiraling trajectories of the perturbing electrons (solid line) and without this allowance (dotted line). The spectra were calculated for the observation at the angle of 45 degrees with respect to the magnetic field. For making the spectra universal, the frequency detuning ω was scaled to the characteristic frequency γ of some of the nonlinear processes in the plasmas (details in [98]).

Figure 18 shows the comparison of the spectra of the entire Ly$_\alpha$ line for the example of tokamak EAST with the allowance for the spiraling trajectories of the perturbing electrons (solid line) and without this allowance (dotted line). The plasma parameters correspond to the edge plasmas of tokamak EAST: B = 2 Tesla, N_e = 5 × 10^{13} cm^{-3}, T = 8 eV. The spectra were calculated for the observation at the angle of 20 degrees with respect to the magnetic field, which is the actual angle of the diagnostic window at EAST. The primary effect of the allowance for the spiraling trajectories of the perturbing electrons is that the ratio of the intensity of the central peak to the intensity of any of the two lateral peaks increased by 150% (from 0.137 to 0.341). The secondary effect is that the shift of the lateral components increased by 11%.

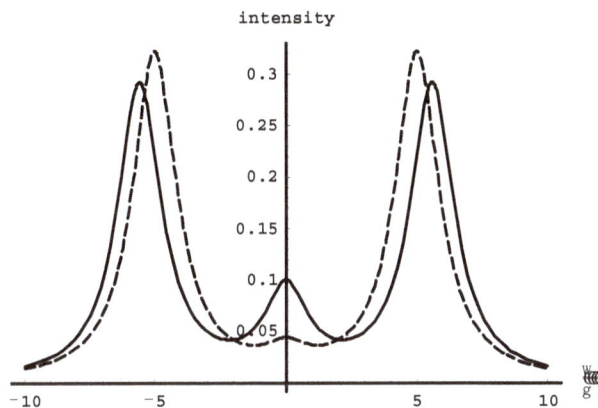

Figure 18. Comparison of the spectra of the Ly$_\alpha$ line for the above example of tokamak EAST with the allowance for the spiraling trajectories of the perturbing electrons (solid line) and without this allowance (dotted line). The spectra were calculated for the observation at the angle of 20 degrees with respect to the magnetic field, which is the actual angle of the diagnostic window at EAST. For making the spectra universal, the frequency detuning ω was scaled to the characteristic frequency γ of some of the nonlinear processes in plasmas (details in [98]).

As the next example, the author of [98] considered parameters that are relevant to the DA white dwarfs (i.e., the white dwarfs emitting hydrogen lines) [99]: B = 100 Tesla, N_e = 10^{17} cm^{-3}, T_e = 5 eV. In this situation, the ratio is 2a/d = 0.4. The primary effect of the allowance for the spiraling trajectories of the perturbing electrons is that the ratio of the intensity of the central peak to the intensity of any of the two lateral peaks increased by 23% (from 1 to 1.23). The secondary effect is that the shift of the lateral (σ-) components increased by 8%.

Thus, both for magnetic fusion plasmas and plasmas of DA white dwarfs (which are physically very different objects), the allowance for the spiraling trajectories of perturbing electrons can play a significant role. In particular, the ratio of the intensity of the central peak to the intensity of any of the two lateral peaks could increase by up to 60%.

10. Influence of Magnetic-Field-Caused Modifications of Trajectories of Plasma Electrons on Shifts and Relative Intensities of Zeeman Components of Hydrogen/Deuterium Spectral Lines: Applications to Magnetic Fusion and White Dwarfs

10.1. Preamble

In paper [98], as reviewed in the above Section 9, there was developed a general framework for calculating shapes of Hydrogen or Deuterium Spectral Lines (HDSL) in strongly-magnetized plasmas with the allowance for helical trajectories of perturbing electrons. The focus of paper [98] was at the relative intensities of the Zeeman components and on their shifts.

In paper [101], the author considered the *direct effect* of helical trajectories of perturbing electrons *on the width* of HDSL. The focus was at the case of a strong magnetic field B, such that the non-adiabatic Stark width practically vanishes and only the adiabatic Stark width remains. Such strong magnetic fields encountered, e.g., in white dwarfs. There was calculated analytically the adiabatic Stark width for this case and compared with the adiabatic Stark width for the rectilinear trajectories of perturbing electrons, the latter being relevant to the case of vanishingly small magnetic fields. It was shown that the adiabatic Stark width calculated with the allowance for helical trajectories of perturbing electrons does not depend on the magnetic field if the magnetic field is sufficiently strong.

In paper [101] it was demonstrated that, depending on the particular HDSL and on plasma parameters, the adiabatic Stark width, as calculated with the allowance for helical trajectories of perturbing electrons, can be either *by orders of magnitude smaller*, or of the same order, or several times higher than the adiabatic Stark width, calculated for rectilinear trajectories of perturbing electrons. It was shown that for the range of plasma parameters typical for DA white dwarfs (i.e., for white dwarfs emitting hydrogen lines), the neglect for the actual, helical trajectories of perturbing electrons can lead to the overestimation of the Stark width by up to one order of magnitude for the alpha- and beta-lines of the Lyman and Balmer series, or to the underestimation of the Stark width by several times for the delta- and higher-lines of the Balmer series. Therefore, the results from paper [101] should motivate astrophysicists for a very significant revision of all the existing calculations of the broadening of various hydrogen lines in DA white dwarfs.

It was also explained in paper [101] that experimental/observational studies, for which the effect of helical trajectories of perturbing electrons on the Stark width might be significant, are not limited by white dwarfs, but can be performed in a variety of laboratory and astrophysical plasmas emitting the hydrogen or deuterium Ly-alpha line.

Below, we reproduce the results from paper [101].

10.2. Analytical Results

For hydrogen/deuterium atoms in a strongly magnetized plasma, the radius-vector $\mathbf{R}(t)$ of a perturbing electron and the electric field $\mathbf{E}(t)$ it creates at the location of the radiating atom, can be represented in the form

$$\mathbf{R}(t) = v_z t \mathbf{B}/B + \rho[1 + (r_{Bp}/\rho)\cos(\omega_B t + \varphi)] + \rho\mathbf{x}\mathbf{B}\,[r_{Bp}/(\rho B)]\sin(\omega_B t + \varphi) \qquad (10.1)$$

$$\mathbf{E}(t) = e\mathbf{R}(t)/[R(t)]^3$$

where the z-axis is chosen along the magnetic field $\mathbf{B}$; $\rho\mathbf{x}\mathbf{B}$ stands for the cross-product (also known as the vector product) of the impact parameter vector ρ and the magnetic field $\mathbf{B}$; e is the electron charge. Here,

$$r_{Bp} = v_p/\omega_B, \quad \omega_B = eB/(m_e c) \qquad (10.2)$$

where v_p is the electron velocity in the plane perpendicular to $\mathbf{B}$; ω_B is the Larmor frequency (also known as the cyclotron frequency).

Without the allowance for helical trajectories of perturbing electrons, the effect of the magnetic field on the width of HDSL becomes noticeable where the magnetic field exceeds certain critical value B_{cr}, The author of paper [101] briefly reminded the physical reason for this. In the so-called Conventional Theory (CT) of the Stark broadening of HDSL (which is frequently referred to as Griem's theory—as presented in Kepple-Griem paper [26] and in Griem's book [13]) the electron impact broadening operator Φ_{ab} (where a and b label, respectively, the upper and lower states that are involved in the radiative transition) is a linear function of the following integral over impact parameters:

$$\int_0^\infty d\rho/\rho \qquad (10.3)$$

This integral diverges both at large and small ρ, so that in the CT the integration got truncated by some ρ_{min} at the lower limit and by some ρ_{max} at the upper limit. The lower limit ρ_{min} is chosen such as to preserve the unitarity of S-matrices involved in the calculation and in then called Weisskopf radius ρ_{We} (discussed in more detail below). At the absence of the magnetic field, the upper limit ρ_{max} is chosen from the requirement that the characteristic frequency v/ρ of the variation of the electric field of the perturbing ion should exceed the plasma electron frequency ω_{pe} according to the general plasma property to screen electric fields at frequencies that were lower than ω_{pe}. So, it is chosen $\rho_{max} = v/\omega_{pe}$, which after substituting the perturbing electron velocity v by the mean thermal velocity v_T, becomes the Debye radius ρ_D.

The uniform magnetic field **B** reduces the spherical symmetry of the problem to the axial symmetry. The fundamental consequence is that in this situation the electron broadening operator Φ_{ab} (and the corresponding Stark width) should be subdivided into two distinct parts: the *adiabatic* part $\Phi_{ad,ab}$ and the *nonadiabatic* part $\Phi_{na,ab}$. The adiabatic part is controlled by the component of the electric field (of the perturbing electron) parallel to **B**, while the nonadiabatic part is controlled by the component of the electric field (of the perturbing electron) perpendicular to **B**.

Physically, the nonadiabatic contribution to the broadening is due to the virtual transitions between the adjacent Zeeman sublevel separated by $\Delta\omega_B = \omega_B/2$. For the nonadiabatic contribution to be effective, the characteristic frequency v/ρ of the variation of the electric field of the perturbing ion should exceed $\Delta\omega_B$. This leads to the modified upper cutoff for the integral (10.3):

$$\rho_{max} = \min(v/\omega_{pe}, v/\Delta\omega_B) = \min(v/\omega_{pe}, 2v/\omega_B) \tag{10.4}$$

which becomes smaller than v/ω_{pe} when

$$\Delta\omega_B = \omega_B/2 > \omega_{pe} \tag{10.5}$$

or

$$B > B_{threshold} = 4c(m_e N_e)^{1/2}, \; B_{threshold}(\text{Tesla}) = 3.62 \times 10^{-7} \, [N_e(\text{cm}^{-3})]^{1/2} \tag{10.6}$$

where N_e is the electron density. Obviously, the greater the ratio B/B_{cr}, the smaller the nonadiabatic contribution to the width becomes. Such inhibition of the nonadiabatic contribution to the width was studied in detail in paper [23]—see also Section 4 of review [102].

In distinction, the adiabatic contribution to the width is not affected by the fulfilment of condition (10.5) or (10.6). Physically this is because the adiabatic contribution is not related to the quantum effect of the virtual transitions between the adjacent Zeeman sublevels. Rather, the physics behind the adiabatic contribution is the phase modulation of the atomic oscillator by the parallel to **B** component of the electric field of the perturbing electron (the phase modulation being, in essence, a classical effect—see. e.g., review [8]).

Thus, it is clear that, as $B/B_{threshold}$ becomes much greater than unity, practically the entire width becomes due to the adiabatic contribution. This happens when $\rho_{max} = v/\Delta\omega_B$ diminishes to the value significantly below $\rho_{min} = \rho_{We}$, i.e., when $\Delta\omega_B > v/\rho_{We}$, which after substituting v by $v_T = (2T_e/m_e)^{1/2}$, T_e being the electron temperature, becomes

$$\Delta\omega B > \Omega We = vT/\rho We \tag{10.7}$$

where Ω_{We} is called the Weiskopf frequency, or

$$B > B_{cr} = 2m_e c v_T/(e\rho_{We}) \tag{10.8}$$

The focus of paper [101] was at very strong magnetic fields that were satisfying the condition (10.8), which will be reformulated more explicitly below. In this situation, the effect of helical trajectories of the perturbing electrons can be presented in the purest form—without the interplay with the magnetic-field-caused inhibition of the nonadiabatic contribution studied in paper [23].

For the purpose of the comparison, in paper [101], first there was calculated the adiabatic contribution for rectilinear trajectories of perturbing electrons in the case of a vanishingly small magnetic field **B**—in the spirit of the CT, but slightly more accurately. By using the parabolic coordinated with the z-axis along **B**, the adiabatic part $(\Phi_{ad,rec})_{ab}$ of the electron broadening operator can be represented in the form (the suffix "rec" stands for "rectilinear trajectories")

$$(\Phi_{ad,rec})_{ab} = -\left[\frac{2h^2 N_e}{3me^2}\right]\left[\frac{(Z_a - Z_b)(Z_a - Z_b)}{a_0{}^2}\right]2\pi < v\ (1/2 + \int_{\rho_{min(v)}}^{\rho_{max(v)}} d\rho/\rho)/v^2 >_{vel} \qquad (10.9)$$

where $< \ldots >_{vel}$ stands for averaging over the distribution of velocities of perturbing electrons and $1/2$ stands for the contribution of the so-called strong collisions (i.e., the collisions with the impact parameters $\rho < \rho_{min}(v)$. In Equation (10.9), Z is the operator of the z-projection of the radius-vector of the atomic electron. In the parabolic coordinates, this operator has only diagonal matrix elements in the manifold of the fixed principal quantum number n, so that the operator $(\Phi_{ad,rec})_{ab}$ also has only diagonal matrix elements, which we denote $_{\alpha\beta}(\Phi_{ad,rec})_{\beta\alpha}$. Here, α and β correspond, respectively, to upper and lower sublevels of the levels a and b that are involved in the radiative transition.

Then, the adiabatic width $\gamma_{\alpha\beta,rec} = -Re[_{\alpha\beta}(\Phi_{ad,rec})_{\beta\alpha}]$ can be expressed, as follows

$$\gamma_{\alpha\beta,rec} = \left[\frac{3h^2 N_e X_{\alpha\beta}{}^2}{2m_e{}^2}\right]2\pi < v\ (1/2 + \int_{\rho_{min(v)}}^{\rho_{max(v)}} d\rho/\rho)/v^2 >_{vel}, \qquad (10.10)$$

where

$$X_{\alpha\beta} = n_a q_\alpha - n_b q_\beta, q_\alpha = (n_1 - n_2)_\alpha, q_\alpha = (n_1 - n_2)_\alpha, \qquad (10.11)$$

where n_1 and n_2 are the parabolic quantum numbers (while q is often called the electric quantum number). Here, $\gamma_{\alpha\beta,rec}$ is the half-width at half maximum of the corresponding component of the line. As for the quantity $X_{\alpha\beta}$, it is the standard label of Stark components of HDSL, but for avoiding any confusion, it was emphasized that in paper [101] there was considered the Zeeman triplet of HDSL, consisting of the central (unshifted) π-component and two σ-components symmetrically shifted to the red and blue parts of the line profile.

For the adiabatic width, one has $\rho_{max}(v) = v/\omega_{pe}$, as explained above. As for $\rho_{min}(v)$, its role is played by the *adiabatic Weisskopf radius*

$$\rho_{Wad}(v) = k_{ad}/v, k_{ad} = 3|X_{\alpha\beta}|\hbar/m_e \qquad (10.12)$$

The adiabatic Weisskopf radius arises naturally without any uncertainty as a result of the *exact* calculation of the adiabatic Stark width in paper [86]. This is the primary distinction of the adiabatic Weisskopf radius from the Weisskopf radius ρ_{WG} in Griem's theory. In the latter, it was chosen from the requirement to preserve the unitarity of the S-matrices entering the calculation:

$$|1 - S_a(\rho, v) S_b{}^*(\rho, v)| \le 2 \qquad (10.13)$$

where the symbol * stands for the complex conjugation.

Therefore, the choice of ρ_{WG} in Griem's theory

$$\rho_{WG}(v) = k_G/v, k_G = \hbar(n_a{}^2 - n_b{}^2)/m_e \qquad (10.14)$$

has an inherent uncertainty by a factor of 2 (for the detailed discussion of this uncertainty see, e.g., paper [103] or Appendix B from book [7]). Below, while calculating the velocity average in Equation (10.9), there was used

$$\rho_{min}(v) = \rho_W(v) = k/v \qquad (10.15)$$

where k would be either k_{ad} from Equation (10.11) or k_G from Equation (10.13), so that for the ratio $\rho_{max}(v)/\rho_{min}(v)$ one has

$$\rho_{max}(v)/\rho_{min}(v) = v^2/(k\omega_{pe}) \qquad (10.16)$$

Then, the result of the integration over the impact parameters in Equation (10.10) can be expressed, as follows

$$\ln[(v^2/v_T^2)/D], D = k\omega_{pe}/v_T^2 \qquad (10.17)$$

While using $k = k_{ad}$ defined in Equation (10.12), the dimensionless constant D becomes

$$D = 3|X_{\alpha\beta}|\hbar\omega_{pe}/(m_e v_T^2) = 3|X_{\alpha\beta}|\hbar\omega_{pe}/(2T_e) = 5.57 \times 10^{-11}|X_{\alpha\beta}|[N_e(\text{cm}^{-3})]^{1/2}/T_e(\text{eV}) \qquad (10.18)$$

At this point, the author of paper [101] calculated the velocity average in Equation (10.10) by employing the three-dimensional isotropic Maxwell distribution $f_3(p)dp = (4/\pi^{1/2})p^2\exp(-p^2)$, where $p = v/v_T$:

$$<\{\ln[(v^2/v_T^2)/D] + 1/2\}/v> _{vel} = (2/\pi^{1/2})[\ln(1/D) + 1/2 - \Gamma] = (2/\pi^{1/2})[\ln(1/D) - 0.0772] \qquad (10.19)$$

where $\Gamma = 0.5772$ is the Euler constant. So, finally, the adiabatic width in the case of rectilinear trajectories becomes:

$$\gamma_{\alpha\beta,rec} = [6\pi^{1/2}\hbar^2 X_{\alpha\beta}^2 N_e/(m_e^2 v_T)] [\ln(1/D) - 0.0772], v_T = (2T_e/m_e)^{1/2} \qquad (10.20)$$

where D was defined in Equation (10.18).

Then, the author of paper [101] started calculating the adiabatic width $\gamma_{\alpha\beta,hel}$ for the case of the actual trajectories (i.e., helical trajectories) of perturbing electrons, the suffix "hel" standing for "helical trajectories". Again, there were considered strong magnetic fields satisfying the condition (10.8). By using the explicit expression (10.12) for the adiabatic Weisskopf radius and by substituting v by v_T, the condition (10.8) can be reformulated more explicitly, as follows

$$B > B_{cr}, B_{cr} = 2T_e/(3|X_{\alpha\beta}|e\lambda_c) \text{ or } B_{cr}(\text{Tesla}) = 9.2 \times 10^2 T_e(\text{eV})/|X_{\alpha\beta}| \text{ Tesla} \qquad (10.21)$$

where $\lambda_c = \hbar/(m_e c) = 2.426 \times 10^{-10}$ cm is the Compton wavelength of electrons.

Such strong magnetic fields are encountered, for example, in plasmas of DA white dwarfs (i.e., white dwarfs emitting hydrogen lines). According to observations, the magnetic field in plasmas of white dwarfs can range from 10^3 Tesla to 10^5 Tesla (see, e.g., papers [104,105]), thus easily exceeding the critical value from Equation (10.21)—given that $T_e \sim 1$ eV in the white dwarfs plasmas emitting hydrogen lines.

For completeness, it should be mentioned that, in paper [101], there was considered the situation where the temperature T_a of the radiating atoms satisfies the condition

$$T_a << (11.12 \text{ keV}/n)(M/M_H)^2 \qquad (10.22)$$

where M is the mass of the radiating atom, M_H is the mass of hydrogen atoms, and n is the principal quantum number of the energy level, from which the spectral line originates. Under this condition, the Lorentz field effects can be disregarded compared to the "pure" magnetic field effects.

It was also noted in paper [101] that *at any value of the magnetic field* (no matter how large or small), the Stark width of the central (unshifted) component of the Ly-alpha Zeeman triplet has practically only the adiabatic contribution—because the non-adiabatic contribution vanishes within the accuracy of about 1%, as shown in detail in paper [23]. So, the experimental/observational studies, for which the effect of helical trajectories of perturbing electrons on the Stark width might be significant, are not limited by white dwarfs, but they can be performed in a variety of laboratory and astrophysical plasmas emitting the hydrogen or deuterium Ly-alpha line (by using the polarization analysis).

Based on Equation (10.1), the z-component (i.e., the component parallel to **B**) of the electric field of the perturbing electron at the location of the radiating atom can be represented in the form

$$E_z(t) = e(v_z t)/[\rho^2 + v_z^2 t^2 + v_p^2/\omega_B^2 + 2(\rho v_p/\omega_B)\cos(\omega_B t + \varphi)]^{3/2} \qquad (10.23)$$

For comparison, in the case of the rectilinear trajectories for vanishingly small **B** it was

$$E_{z,rec}(t) = e(\rho e_z + v e_z t)/(\rho^2 + v^2 t^2)^{3/2} \qquad (10.24)$$

where e_z is the unit vector along the z-axis. Here and below, for any two vectors **a** and **b**, the notation **ab** stands for their scalar product (also known as the dot-product). It should be noted that Equation (10.23) in the limit of B = 0 does not reduce to Equation (10.24). This is because Equation (10.23) adequately describes the situation only for the strong magnetic fields defined by the condition

$$(v_p/\omega_B)^2 \ll [\rho_{Wad}(v_z)]^2 \qquad (10.25)$$

It can be reformulated as

$$[m_e/(3X_{\alpha\beta}e\lambda_c B)]^2 v_p^2 v_z^2 \ll 1 \qquad (10.26)$$

which after substituting v_p^2 by its average value over the two-dimensional Maxwell distribution $<v_p^2>$ = v_T^2 and v_z^2 by its average value over the one-dimensional Maxwell distribution $<v_z^2>$ = $v_T^2/2$, can be rewritten as

$$B \gg B_{min} = 2^{1/2} m_e v_T^2/(6|X_{\alpha\beta}|e\lambda_c) = 2^{1/2} T_e/(3|X_{\alpha\beta}|e\lambda_c) \text{ or } B_{min}(Tesla) = 6.5 \times 10^2 T_e(eV)/|X_{\alpha\beta}| \qquad (10.27)$$

This condition is similar to the condition (10.8) or (10.21), under which practically the entire width of HDSL becomes due to the adiabatic contribution only.

The starting formula for the adiabatic width in the case of helical trajectories of perturbing electrons $\gamma_{\alpha\beta,hel} = -\text{Re}[_{\alpha\beta}(\Phi_{ad,hel})_{\beta\alpha}]$ (the suffix "hel" stands for "helical trajectories")

$$\gamma_{\alpha\beta,hel} = N_e \int_{-\infty\rightarrow}^{\infty\rightarrow} dV_z f_1(V_z) \int_0^\infty dV_p f_1(V_p)\left(V_z^2 + V_p^2\right)^{\frac{1}{2}} \sigma(V_z, V_p), \qquad (10.28)$$

where the operator $\sigma(v_z, v_p)$, which is physically the cross-section of the so-called optical collisions that are responsible for the broadening of the spectral line, has the form:

$$\sigma(V_z, V_p) = \int_0^\infty d\rho\, 2\pi\rho \left[1 - S_a(, V_z, V_p) S_b^*(, V_z, V_p)\right]_\varphi \qquad (10.29)$$

Here, $f_1(v)$ and $f_2(v)$ are the 1D- and 2D-Maxwell distributions, respectively; S_a and S_b are the corresponding scattering matrices; the symbol $[\dots]_\varphi$ stands for the average over the phase φ entering Equation (10.23). In the spirit of the CT, one gets

$$1 - S_a(, V_z, V_p) S_b * (, V_z, V_p) = [3X_{\alpha\beta}h/(2m_e e)]^2 \int_{-\infty}^\infty dt E_z(t) \int_{-\infty}^t dt_1 E_z(t_1), \qquad (10.30)$$

where $E_z(t)$ is given by Equation (10.23).

It is important to emphasize the following. The double integral in Equation (10.30) vanishes for the odd part of Ez(t), while for the even part of Ez(t) it is equal to the one half of the square of

$$\int_{-\infty}^{\infty} dtE_z(t)$$

Therefore it is sufficient to calculate the double integral just for the even part of Ez(t), i.e., for

$$E_z(t)_{even} = [E_z(t) - E_z(-t)]/2 \tag{10.31}$$

Before doing this, the author of paper [101] broke the integration over the impart parameters in Equation (10.29) into the following two parts

$$\sigma(v_z, v_p) = \sigma_1(v_z, v_p) + \sigma_2(v_z, v_p) \tag{10.32}$$

where

$$\sigma_1(v_z, v_p) = \int_{\rho_0}^{\infty} d\rho \, 2\pi\rho \left[1 - S_a(\rho, v_z, v_p) S_b{}^*(\rho, v_z, v_p)\right]_\varphi \tag{10.33}$$

$$\sigma_2(v_z, v_p) = \int_0^{\rho_0} d\rho \, 2\pi\rho \left[1 - S_a(\rho, v_z, v_p) S_b{}^*(\rho, v_z, v_p)\right]_\varphi \tag{10.34}$$

Here

$$\rho_0 = v_p/\omega_B \tag{10.35}$$

While expanding Ez(t)$_{even}$ from Equation (10.31), in terms of the small parameter $\rho_0/\rho = v_p/(\omega_B\rho)$ and keeping the first non-vanishing term of the expansion, one obtains

$$E_z(t)_{even} = (\sin\varphi)(3e\rho v_p v_z/\omega_B) \, t[\sin(\omega_B t)]/(\rho^2 + v_z^2 t^2)^{5/2} \tag{10.36}$$

Then,

$$\int_{-\infty}^{\infty} dtE_z(t)_{even} \int_{-\infty}^t dt_1 E_z(t_1)_{even} = \left(\frac{1}{2}\right)\left[\int_{-\infty}^{\infty} dtE_z(t)\right]^2 = (\sin^2\varphi)\left(\frac{2e^2\omega_B{}^2 v_p{}^2}{v_z{}^6}\right)\left[K_1\left(\frac{\omega_B\rho}{|v_z|}\right)\right]^2 \tag{10.37}$$

where $K_1(s)$ is the modified Bessel function of the second kind. After averaging over the phase φ in Equation (10.37), one gets

$$[1 - S_a(\rho, v_z, v_p) \, S_b{}^*(\rho, v_z, v_p)]_\varphi = [3X_{\alpha\beta}\hbar/(2m_e)]^2 \, (\omega_B{}^2 v_p{}^2/v_z{}^6)[K_1(\omega_B\rho/|v_z|)]^2 \tag{10.38}$$

After substituting the expression (10.38) into Equation (10.33) and integrating over the impact parameters, we obtain

$$\sigma_1(v_z, v_p) = (9\pi^{3/2}/8)(X_{\alpha\beta}\hbar/m_e)^2(v_p{}^2/v_z{}^4) \, MeijerG[\{\{\},\{3/2\}\},\{\{0,0,2\},\{\}\}, v_p{}^2/v_z{}^2] \tag{10.39}$$

where MeijerG[...] is the Meijer G-function.

Then, the author of paper [101] proceeded to calculating $\sigma_2(v_z, v_p)$ defined by Equation (10.34). While expanding Ez(t)$_{even}$ from Equation (10.31) in terms of the small parameter $\rho/\rho_0 = \omega_B\rho/v_p$ and keeping the first non-vanishing term of the expansion, one obtains

$$E_z(t)_{even} = (\sin\varphi)(3e\rho v_p v_z/\omega_B) \, t[\sin(\omega_B t)]/(v_p{}^2/\omega_B{}^2 + v_z{}^2 t^2)^{5/2} \tag{10.40}$$

Then,

$$\int_{-\infty}^{\infty} dt E_z(t)_{even} \int_{-\infty}^{t} dt_1 E_z(t_1)_{even} = \left(\tfrac{1}{2}\right)[\int_{-\infty}^{\infty} dt E_z(t)]^2 = (\sin^2 \varphi)\left(\tfrac{2e^2\rho^2\omega_B^4}{v_z^6}\right)\{K_1[\left(\tfrac{v_p^2}{v_z^2}\right)^{1/2}]\}^2, \tag{10.41}$$

After averaging over the phase φ in Equation (10.41), we get

$$[1 - S_a(\rho,v_z, v_p)\, S_b{}^*(\rho, v_z, v_p)]_\varphi = [3X_{\alpha\beta}\hbar/(2m_e)]^2\, (\omega_B^4\rho^2/v_z^6)\{K_1[(v_p^2/v_z^2)^{1/2}]\}^2 \tag{10.42}$$

After substituting the expression (10.42) into Equation (10.34) and integrating over the impact parameters, one obtains

$$\sigma_2(v_z, v_p) = (9\pi/8)\,(X_{\alpha\beta}\hbar/m_e)^2(v_p^4/v_z^6)\,\{K_1[(v_p^2/v_z^2)^{1/2}]\}^2 \tag{10.43}$$

By combining Equations (10.39) and (10.43), one gets:

$$\sigma(v_z, v_p) = \sigma_1(v_z, v_p) + \sigma_2(v_z, v_p) =$$
$$(9\pi/8)(X_{\alpha\beta}\hbar/m_e)^2(v_p^2/v_z^4)\{\pi^{1/2}\text{MeijerG}[\{\{\},\{3/2\}\},\{\{0,0,2\},\{\}\},v_p^2/v_z^2]+(v_p/v_z)^2 \tag{10.44}$$
$$K_1[(v_p^2/v_z^2)^{1/2}]^2\}$$

According to Equation (10.28), the adiabatic width in the case of helical trajectories of perturbing electrons is

$$\gamma_{\alpha\beta,hel} = N_e <(v_z^2 + v_p^2)^{1/2}\, \sigma(v_z, v_p)>_{veloc} \tag{10.45}$$

where $<\dots>_{veloc}$ stands for the average over velocities v_p and v_z. In order to get the message across in a relatively uncomplicated form, the author of paper [101] simply substituted v_p^2 by its average value over the two-dimensional Maxwell distribution $<v_p^2> = v_T^2$ and v_z^2 by its average value over the one-dimensional Maxwell distribution $<v_z^2> = v_T^2/2$. As a result, he obtained the final expression:

$$\gamma_{\alpha\beta,hel} = 7.9(X_{\alpha\beta}\hbar/m_e)^2 N_e/v_T \text{ or } \gamma_{\alpha\beta,hel}(s^{-1}) = 1.8 \times 10^{-7} X_{\alpha\beta}^2[N_e(cm^{-3})]/[T_e(eV)]^{1/2}, v_T = (2T_e/m_e)^{1/2} \tag{10.46}$$

It should be noted that $\gamma_{\alpha\beta,hel}$ does not depend on the magnetic field in the case where the magnetic field is strong enough to satisfy the condition (10.27).

The role of the allowance for helical trajectories of perturbing electrons can be best understood by considering the ratio of $\gamma_{\alpha\beta,hel}$ from Equation (10.46) to $\gamma_{\alpha\beta,rec}$ from Equation (10.20):

$$ratio = \gamma_{\alpha\beta,hel}/\gamma_{\alpha\beta,hel} = 0.74/[\ln(1/D) - 0.0772] \tag{10.47}$$

Figure 19 shows this ratio versus the dimensionless parameter D, which is defined by Equation (10.18) and that is physically the ratio $\rho_{Wad}(v_T)/\rho_D$. It is seen that for D < 0.44, the allowance for helical trajectories of perturbing electrons *decreases* the adiabatic width, while for D > 0.44, the allowance for helical trajectories of perturbing electrons *increases* the adiabatic width. The fact that, the allowance for helical trajectories of perturbing electrons could lead to two different outcomes (i.e., to either decreasing or increasing the adiabatic width of HDSL) is a *counterintuitive result*.

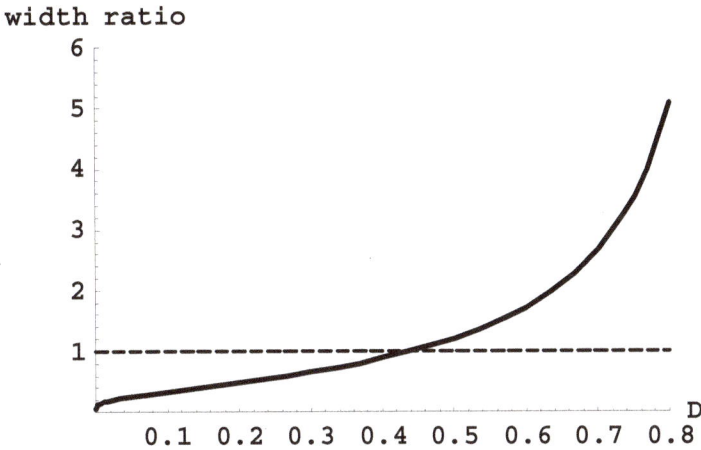

Figure 19. The ratio of adiabatic widths $\gamma_{\alpha\beta,\mathrm{hel}}/\gamma_{\alpha\beta,\mathrm{hel}}$ versus the dimensionless parameter D, which is defined by Equation (10.18) and which is physically the ratio $\rho_{\mathrm{Wad}}(v_T)/\rho_D$ (solid curve). The dashed horizontal line is there for better visualizing the two regions: $\gamma_{\alpha\beta,\mathrm{hel}}/\gamma_{\alpha\beta,\mathrm{hel}} < 1$ and $\gamma_{\alpha\beta,\mathrm{hel}}/\gamma_{\alpha\beta,\mathrm{hel}} > 1$ [101].

Figure 20 shows the ratio $\gamma_{\alpha\beta,\mathrm{hel}}/\gamma_{\alpha\beta,\mathrm{hel}}$ from Equation (10.47) versus the electron density N_e for the range of N_e relevant to the DA white dwarfs at $T_e = 1$ eV. The lower curve is for the Lyman-alpha line, for which the adiabatic width is non-zero only for the two π-components of $|X_{\alpha\beta}| = 2$. The upper curve is for the Balmer-beta line—specifically for its two intense π-components of $|X_{\alpha\beta}| = 10$.

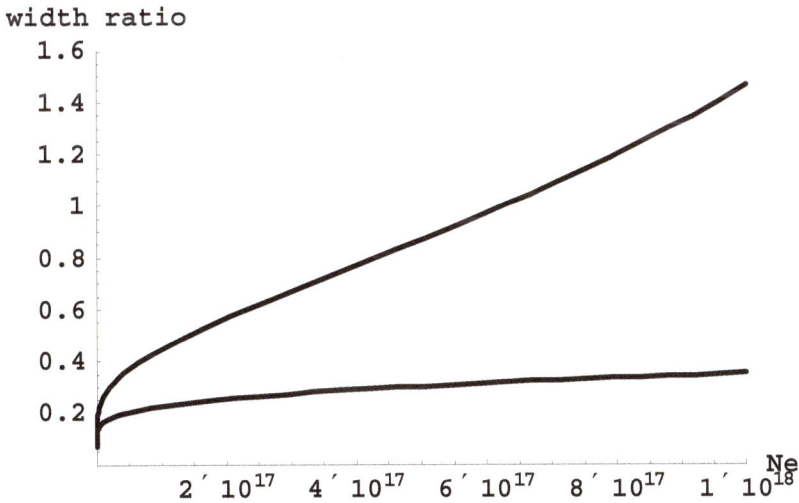

Figure 20. The ratio $\gamma_{\alpha\beta,\mathrm{hel}}/\gamma_{\alpha\beta,\mathrm{hel}}$ from Equation (10.47) versus the electron density N_e for the range of N_e relevant to the DA white dwarfs at $T_e = 1$ eV. The lower curve is for the Lyman-alpha line. The upper curve is for the Balmer-beta line—specifically for its two intense π-components of $|X_{\alpha\beta}| = 10$ [101].

Figure 21 shows more clearly the ratio $\gamma_{\alpha\beta,\mathrm{hel}}/\gamma_{\alpha\beta,\mathrm{hel}}$ for the electron densities below 2×10^{17} cm^{-3} for the Lyman-alpha line (the lower curve) and for the two intense π-components of $|X_{\alpha\beta}| = 10$ of the Balmer-beta line (the middle curve). The upper curve shows the ratio $\gamma_{\alpha\beta,\text{hel}}/\gamma_{\alpha\beta,\text{hel}}$ for the two most intense π-components of $|X_{\alpha\beta}| = 28$ of the Balmer-delta line.

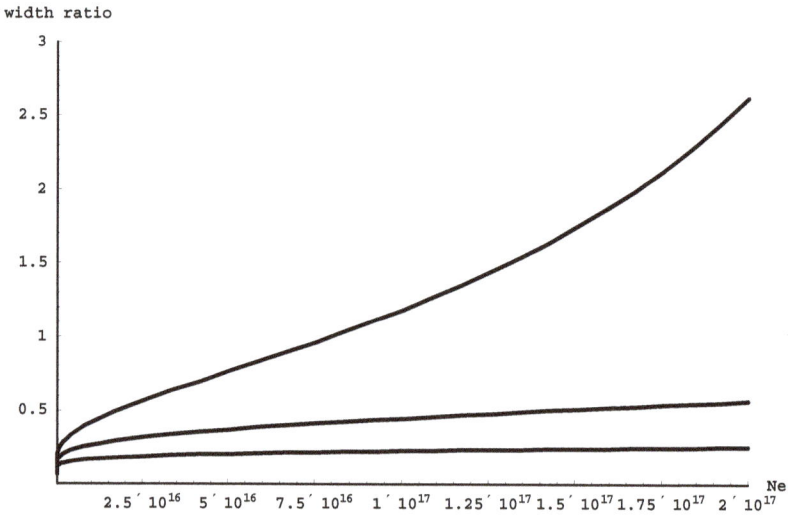

Figure 21. The ratio $\gamma_{\alpha\beta,\text{hel}}/\gamma_{\alpha\beta,\text{hel}}$ for the electron densities below 2×10^{17} cm^{-3} for the Lyman-alpha line (the lower curve) and for the two intense π-components of $|X_{\alpha\beta}| = 10$ of the Balmer-beta line (the middle curve). The upper curve shows the ratio $\gamma_{\alpha\beta,\text{hel}}/\gamma_{\alpha\beta,\text{hel}}$ for the two most intense π-components of $|X_{\alpha\beta}| = 28$ of the Balmer-delta line [101].

In paper [101], there were also presented explicit practical formulas for the adiabatic width (with the allowance for helical trajectories of perturbing electrons in the case of strong magnetic fields satisfying the condition (10.27)) in the wavelength scale for the following five HDSL—namely, the Full Width at Half Maximum (FWHM) $\Delta\lambda_{1/2,\text{ad}}$.

For Lyman-alpha:

$$\Delta\lambda_{1/2,\text{ad}}(\text{nm}) = 1.1 \times 10^{-20} N_e(\text{cm}^{-3})/[T_e(\text{eV})]^{1/2} \tag{10.48}$$

For Balmer-alpha line components:

$$\Delta\lambda_{1/2,\text{ad}}(\text{nm}) = 8.1 \times 10^{-20} X_{\alpha\beta}{}^2 N_e(\text{cm}^{-3})/[T_e(\text{eV})]^{1/2} \tag{10.49}$$

For Balmer-beta line components:

$$\Delta\lambda_{1/2,\text{ad}}(\text{nm}) = 4.5 \times 10^{-20} X_{\alpha\beta}{}^2 N_e(\text{cm}^{-3})/[T_e(\text{eV})]^{1/2} \tag{10.50}$$

For Balmer-gamma line components:

$$\Delta\lambda_{1/2,\text{ad}}(\text{nm}) = 3.7 \times 10^{-20} X_{\alpha\beta}{}^2 N_e(\text{cm}^{-3})/[T_e(\text{eV})]^{1/2} \tag{10.51}$$

For Balmer-delta line components:

$$\Delta\lambda_{1/2,\text{ad}}(\text{nm}) = 3.2 \times 10^{-20} X_{\alpha\beta}{}^2 N_e(\text{cm}^{-3})/[T_e(\text{eV})]^{1/2} \tag{10.52}$$

It should be reminded that $X_{\alpha\beta}$ is the combination of the parabolic quantum numbers that are defined in Equation (10.11).

Finally, it should be noted that under the condition (10.21), the Stark profile of the entire HDSL is simply the sum of Lorentzians corresponding to each Stark component

$$S(\Delta\omega) = \Sigma j_{\alpha\beta} L_{\alpha\beta}(\Delta\omega),$$
$$L_{\alpha\beta}(\Delta\omega) = \int\limits_0^\infty dFW(F)\left(\frac{1}{\pi}\right)\gamma_{\alpha\beta}/[\gamma_{\alpha\beta}^2 + (\Delta\omega - C_{\alpha\beta}F)^2], \ \gamma_{\alpha\beta} = \gamma_{\alpha\beta,hel} + \gamma_{\alpha\beta,nat}, \rightarrow C_{\alpha\beta} = 3X_{\alpha\beta}\frac{h}{2m_e e}. \qquad (10.53)$$

Here, $j_{\alpha\beta}$ is the relative intensity of the Stark component labeled "$\alpha\beta$", $W(F)$ is the distribution of the quasistatic field F, $\gamma_{\alpha\beta,nat}$ is the natural (radiative) width of the particular Stark component. The summation is over both π- and σ-components, but for the latter $\gamma_{\alpha\beta,hel} = 0$.

10.3. Comparison with the Existing Simulation

In 2018, Rosato et al. published a brief paper [106] containing just one simulation of the effect of helical trajectories of perturbing electrons on the adiabatic width of the π-component of the Lyman-alpha line only at one set of plasma parameters: B = 2000 Tesla, $N_e = 10^{17}$ cm^{-3}, $T_e = 1$ eV. (There was no reference in their paper on paper [101] of 2017 that provided the analytical solution of the same effect for *any* hydrogen (or deuterium) spectral line for *any* electron density N_e and *any* electron temperature T_e). The simulated FWHM of the π-component of the Lyman-alpha line, as obtained by Rosato et al. [106] for the above one set of plasma parameters was 1.6 pm (as can be deduced from the simulated line profile in their Figure 2a). From the analytical result from paper [106], represented by Equation (10.48) above, the corresponding FWHM is 1.1 pm.

The difference between the simulated FWHM and the analytically calculated FWHM for this spectral line at this set of plasma parameters is about 30%. So, this simulation basically confirmed the above analytical result. As for the reasons for the 30% difference, there could be one of the two or both of the following.

First, B = 2000 Tesla exceeds $B_{min}(T_e=1$ eV, $X_{\alpha\beta}=2) = 325$ Tesla from Equation (10.27) only by six times. In other words, B = 2000 Tesla might not be strong enough for the theoretical result to be very accurate.

Second, (but more probably) there could have been an error in the simulation by Rosato et al. from paper [106]. In the past, there were numerous errors in various simulations by Rosato et al., producing outcomes that contradicted rigorous analytical results by various authors, as demonstrated in review [102] and in paper [98].[5]

There are also conceptual errors in paper [106]. For the values of the parameter D < 0.44 (where $D = 5.57 \times 10^{-11}|X_{\alpha\beta}|[N_e(\text{cm}^{-3})]^{1/2}/T_e(\text{eV})$ according to Equation (10.18)), including the set of $N_e = 10^{17}$ cm^{-3}, $T_e = 1$ eV, $X_{\alpha\beta} = 2$, the allowance for the helical trajectories of perturbing electrons decreases the adiabatic width when compared to the corresponding result for rectilinear trajectories, as it was shown analytically in paper [101] in 2017. When Rosato et al. [106] in 2018 tried to give a physical explanation of this kind of the result of their simulation (while claiming erroneously that they were the first to discover this effect), they referred to the upper cutoff (for the integration over impact parameters) $\rho_{max} = \min(v/\omega_{pe}, v/\Delta\omega_B) = v/\Delta\omega_B$ in case of $\Delta\omega_B > \omega_{pe}$. However, in reality, this cutoff is effective only for the non-adiabatic contribution to the width, but has no effect on the adiabatic contribution to the width, the latter being the subject of their paper [106] (as well as of paper [101]).

Another conceptual error by Rosato et al. [106] was to imply that in strongly magnetized plasmas, the allowance for helical trajectories of perturbing electrons always decreases the adiabatic width, regardless of N_e and T_e. However, in reality for the values of the parameter

---

[5] For example, simulations by Rosato et al. [107] yielded a non-zero impact shift of the σ-components of the Ly$_\alpha$ line (for rectilinear trajectories of the perturbers)—contrary to the well-known rigorous analytical result [108]. Also, Rosato et al. [107] overestimated the primary, adiabatic contribution to the dynamical Stark broadening by ions in magnetic fusion plasmas by up to an order of magnitude [98,102]. In addition, for the Ly$_\alpha$ line, simulations by Rosato et al. [108] yielded an infinite result for the broadening function $A_-(\rho)$ at $\rho = 0$—contrary to the rigorous analytical result $A_-(\rho) = \text{const} \ \rho = 0$ at $\rho = 0$ [98,102].

D > 0.44, the adiabatic width is greater than if it were calculated for rectilinear trajectories of perturbing electrons. This conceptual error by Rosato et al. [106] stems from the general inferiority of simulations as compared to the analytical theory: simulations are unable to provide the functional dependence of the effect under consideration on various input parameters (in distinction to the analytical theory), and thus often miss the big picture—see, e.g., [9,10].

There was yet another conceptual error by Rosato et al. [106]. They wrote: "The Zeeman central component of Lyman α has been considered within an adiabatic model, suitable for strong magnetic field regimes where interactions between Δm = ±1 levels are negligible". However, in reality the Zeeman central component of the Lyman-alpha line has practically only the adiabatic width *not just for strong magnetic fields, but for any magnetic field*—because the non-adiabatic contribution to the width of this component vanishes within the accuracy of 1%, as it was rigorously shown analytically in paper [23].

10.4. Closing Remarks

In paper [101], there was considered the effect of helical trajectories of perturbing electrons on the width of HDSL for the case of strong magnetic fields, such that the non-adiabatic Stark width practically vanishes and only the adiabatic Stark width remains. Such strong magnetic fields encountered, e.g., in white dwarfs. There was calculated analytically the adiabatic Stark width for this case and its ratio to the adiabatic Stark width for the rectilinear trajectories of perturbing electrons. It was demonstrated that the adiabatic Stark width calculated with the allowance for helical trajectories of perturbing electrons does not depend on the magnetic field for the case of strong magnetic fields under consideration.

It was shown in paper [101] that, depending on the particular HDSL and on plasma parameters, the adiabatic Stark width, calculated with the allowance for helical trajectories of perturbing electrons, can be either *by orders of magnitude smaller*, or of the same order, or several times higher than the adiabatic Stark width, as calculated for rectilinear trajectories of perturbing electrons. Such a variety of outcomes is a counterintuitive result. It was also demonstrated that for the range of plasma parameters typical for DA white dwarfs (i.e., for white dwarfs emitting hydrogen lines), the neglect for the actual, helical trajectories of perturbing electrons can lead to:

- the *overestimation of the Stark width by up to one order of magnitude* for the alpha- and beta-lines of the Lyman and Balmer series;
- the *underestimation of the Stark width by several times* for the delta- and higher-lines of the Balmer series.

Therefore, the results from paper [101] should motivate astrophysicists for a *very significant revision* of all the existing calculations of the broadening of hydrogen lines in DA white dwarfs.

The last but not least: *at any value of the magnetic field* (no matter how large or small), the Stark width of the central (unshifted) component of the Ly-alpha Zeeman triplet has practically only the adiabatic contribution, as shown in detail in paper [23]. So, the experimental/observational studies, for which the effect of helical trajectories of perturbing electrons on the Stark width might be significant, are not limited by white dwarfs, but can be performed in a variety of laboratory and astrophysical plasmas emitting the hydrogen or deuterium Ly-alpha line (by using the polarization analysis).

11. Stark Broadening of Hydrogen Lines in Plasmas of Electron Densities up to or More than $N_e \sim 10^{20}$ cm^{-3}

11.1. Preamble

All through the long history of experimental and theoretical studies of the Stark Broadening of Hydrogen Spectral Lines (SBHSL) in plasmas, benchmark experiments (i.e., experiments where plasma parameters were measured independently of the Stark broadening) played a very important role. As a new benchmark experiment was performed at some novel plasma source at the range of the electron densities N_e higher than for the previous benchmark experiment performed at a different plasma source, almost always discrepancies were found with existing theories. In this way, benchmark experiments stimulated developing more advanced theories—the theories allowing for various high-density effects.

The most recent benchmark experiment by Kielkopf and Allard (hereafter, KA) [109], where the SBHSL was tested while using the H_α line, was performed at a laser-produced pure-hydrogen plasma reaching $N_e = 1.4 \times 10^{20}$ cm^{-3}. This exceeded by two orders of magnitude the highest values of $N_e \sim (3\text{--}4) \times 10^{18}$ cm^{-3} reached by the corresponding previous benchmark experiments: by Kunze group (Büscher et al. [110]) at the gas-liner pinch[6] and by Vitel group (Flih et al. [112]) at the flash tube plasma.

At the electron densities reached in the KA experiment [109], no theoretical calculations of the Full Width at Half Maximum (FWHM) of the H_α line existed. Indeed, the highest value of N_e in the tables of FWHM of the H_α line by Gigosos and Cardenoso [113], produced by fully-numerical simulations, was 4.64×10^{18} cm^{-3} (their simulations are considered by the research community as the most advanced). In the frames of the so-called standard (or conventional) analytical theory, Kepple and Griem [28] calculated the FWHM of the H_α line up to $N_e = 10^{19}$ cm^{-3}, because, at higher values of N_e, the standard theory becomes invalid. (The primary distinction between the standard analytical theory [28] and Gigosos-Cardenoso simulations [113] is that the latter allowed for the ion dynamics in distinction to the former; however, the role of the ion dynamics diminishes as N_e increases and it becomes practically insignificant at values of $N_e \sim 10^{19}$ cm^{-3} and higher.) All other simulations and analytical methods (except the one discussed in the next paragraph), reviews of which can be found, e.g., in book [5] and paper [114], listed the FWHM of the H_α line either up to $N_e \sim 4 \times 10^{18}$ cm^{-3} or lower.

As for the theory by Kielkopf and Allard, in their paper [109] extended above $N_e > 10^{20}$ cm^{-3}, it is inconsistent because it completely neglects the contribution of plasma electrons to the Stark broadening. If Kielkopf and Allard would have attempted including the broadening by electrons, their theoretical widths would have increased by about (50–60)%, and thus would have overestimated the experimental widths at $N_e > 10^{19}$ cm^{-3} by about (50–60)%. By the way, at the lowest density of Kielkopf-Allard experiment $N_e = 8.65 \times 10^{17}$ cm^{-3}, the experimental width is by 50% greater than their theoretical width, so that the inclusion of the broadening by electrons would bring their theoretical width in agreement with the experimental width for $N_e < 10^{19}$ cm^{-3}, and is thus in agreement with other theories for $N_e < 10^{19}$ cm^{-3}; this fact further underscores the inconsistent nature of Kielkopf-Allard theory. The contributions of ions and electrons to the broadening are of the same order of magnitude. For the above reason, the "agreement" of their theory with their experimental widths at $N_e > 10^{19}$ cm^{-3} is fortuitous.

Therefore, in paper [115], the results of which we present here, the author developed a consistent analytical theory that is relevant to the range of the electron densities reached in KA experiment [109]. At this range of N_e, a new factor becomes significant for the SBHSL—the factor that is never taken into account in any previous simulations or analytical theories of the SBHSL. This new factor is a rising contribution of the Electrostatic Plasma Turbulence (EPT) at the *thermal* level of its energy density.

The EPT at any level of its energy density is represented by oscillatory electric fields F_t arising when the waves of the separation of charges propagate through plasmas: they correspond to *collective*

---

6   In the earlier experiment at the gas-liner pinch (Böddeker et al. [111]), the densities up to $N_e \sim 10^{19}$ cm^{-3} had been reached. However, the experiment by Böddeker et al. [111] had deficiencies, which were addressed and eliminated in the experiment by Büscher et al. [110]. In distinction to the former experiment, in the latter one: (a) the spectroscopic measurements were performed simultaneously with the diagnostics; (b) highly reproducible discharge condition was used where the H_α line was measured spatially resolved along the discharge axis indicating that no inhomogeneities along the axis existed; and, (c) high care has been taken to prevent the optical thickness.

degrees of freedom in plasmas—in distinction to the electron and ion microfields that correspond to individual degrees of freedom of charged particles. In relatively low density plasmas, various kinds of the EPT at the *supra-thermal* levels of its energy density, specifically at the levels that are several orders of magnitude higher than the thermal level, were discovered experimentally via the enhanced ("anomalous") SBHSL in numerous experiments/observations that were performed by different groups at various plasma sources [90,116–129], some of these experiments being summarized in books [5,7].

As for the EPT at the thermal level of its energy density (hereafter, "thermal EPT"), their contribution to the SBHSL in relatively low density plasmas is by several orders of magnitude smaller than the contribution of the electron and ion microfields, so that their effect was negligibly small, and therefore never detected spectroscopically. However, at the range of the electron densities reached in KA experiment [109], the contribution to the SBHSL from the thermal EPT becomes comparable to the contribution of the electron and ion microfields.

So, in paper [115], the author took into account the contribution to the SBHSL from the thermal EPT. As a result, the theoretical FWHM of the H_α line became in a very good agreement with the experimental FWHM of the H_α line by KA [109] in the entire range of their electron densities, including the highest electron density $N_e = 1.4 \times 10^{20}$ cm^{-3}.

11.2. Theory and the Comparison with the Experiment

According to Bohm and Pines [130], the number of collective degrees of freedom in a unit volume of a plasma is $N_{coll} = 1/(6\pi^2 r_D{}^3)$, where r_D is the Debye radius. Therefore, the energy density of the oscillatory electric fields at the thermal level is $F_t{}^2/(8\pi) = N_{coll}T/2$, so that

$$F_t{}^2 = 16\pi^{1/2}e^3N_e{}^{3/2}/(3T^{1/2}) \tag{11.1}$$

where T is the temperature and e is the electron charge.

At the absence of a magnetic field, there are only two types of the EPT: Langmuir waves/turbulence and ion acoustic waves/turbulence (a.k.a. ionic sound). Langmuir waves are the high-frequency branch of the EPT. Its frequency is approximately the plasma electron frequency

$$\omega_{pe} = (4\pi e^2 N_e/m_e)^{1/2} = 5.64 \times 10^4 \, N_e{}^{1/2} \tag{11.2}$$

Ion acoustic waves are the low-frequency branch of the EPT. They are represented by a broadband oscillatory electric field, whose frequency spectrum is below or of the order of the ion plasma frequency.

$$\omega_{pi} = (4\pi e^2 N_i Z^2/m_i)^{1/2} = 1.32 \times 10^3 \, Z(N_i m_p/m_i)^{1/2} \tag{11.3}$$

where N_i is the ion density, Z is the charge state; m_e, m_p, and m_i are the electron, proton, and ion masses, respectively. In the "practical" parts of Equations (11.2) and (11.3), CGS units are used. Below we set Z = 1, so that $N_i = N_e$.

The thermal energy density of the collective degrees of freedom $E_{tot}{}^2/(8\pi) = N_{coll}T/2$ is distributed in equal parts between the high- and low-frequency branches:

$$F_0{}^2 = E_0{}^2 = E_{tot}{}^2/2 \tag{11.4}$$

where F_0, E_0, and E_{tot} are the root-mean-square (rms) thermal electric fields of the ion acoustic turbulence, the Langmuir turbulence, and the total turbulence, respectively.

In paper [115], there was first discussed the contribution of the thermal ion acoustic turbulence to the SBHSL. It is useful to begin by estimating a ratio of the rms thermal electric field F_0 of the ion acoustic turbulence to the standard characteristic value F_N of the ion microfield, where

$$F_N = 2\pi(4/15)^{2/3}eN_e{}^{2/3} = 2.603 \, eN_e{}^{2/3} = 3.751 \times 10^{-7} \, [N_e(cm^{-3})]^{2/3} \, V/cm \tag{11.5}$$

Using Equations (11.1), (11.4) and (11.5), in paper [115], there was obtained:

$$F_0/F_N = 0.1689 \, N_e^{1/12}/[T(K)]^{1/4} \tag{11.6}$$

where the temperature T is in Kelvin. At the highest density point of KA experiment [109] ($N_e = 1.39 \times 10^{20}$ cm^{-3}, T = 34486 K), Equation (11.6) yields $F_0/F_N = 0.59$. This shows that in the conditions of KA experiment [109], the rms thermal electric field of the ion acoustic turbulence becomes comparable to the standard ion microfield.

In a broad range of plasma parameters, especially at the range of densities of KA experiment [109], radiating hydrogen atoms perceive oscillatory electric fields of the ion acoustic turbulence as quasistatic. In any code that is designed for calculating shapes of spectral lines from plasmas, an important task becomes the averaging over the ensemble distribution W(**F**) of the *total* quasistatic field **F** = **F**$_t$ + **F**$_i$, where **F**$_i$ is the quasistatic part of the ion microfiled (for the range of densities of KA [109] almost the entire ion microfiled is quasistatic). In other words, the key part of the problem becomes the calculation of W(**F**).

This distribution was derived in paper [131] in the following form:

$$W(a, \beta) = [\frac{3}{(2\pi)^{\frac{1}{2}}}]a\beta \int_0^{\infty} du\{\exp[-3a^2(\beta - u)^2] - \exp[-3a^2(\beta + u)^2]\}W_i(u)/u \tag{11.7}$$

where

$$\beta = F/F_N, u = F_i/F_N \tag{11.8}$$

For the distribution of the quasistatic part of the ion microfield $W_i(u)$ in Equation (11.7), in paper [115], there was used the APEX distribution [132].

Then, the author of paper [115] discussed the contribution of the thermal Langmuir turbulence to the SBHSL. The Langmuir turbulence, being the high-frequency one, causes a *dynamical* SBHSL—similar to the dynamical SBHSL by the electron microfield. In paper [133] there was derived analytically the Langmuir-turbulence-caused contribution (additional to the electron microfield contribution) to the real part $\Gamma = -\text{Re } \Phi$ of the dynamical broadening operator Φ. In particular, diagonal elements of Γ have the form

$$\Gamma_{\alpha\beta} = \Gamma_\alpha + \Gamma_\beta - d_{\alpha\alpha} \, d_{\beta\beta} \, E_0^2\gamma_p/[3\hbar^2(\gamma_p^2 + \omega_{pe}^2)] \tag{11.9}$$

where

$$\Gamma_\alpha = [E_0^2\gamma_p/(12\hbar^2)]\{2d_{\alpha\alpha}^2/(\gamma_p^2+\omega_{pe}^2) + (|\, d_{\alpha,\alpha-1}|^2 + |\, d_{\alpha,\alpha+1}|^2)[1/(\gamma_p^2 + (\omega_F - \omega_{pe})^2) + 1/(\gamma_p^2 + (\omega_F + \omega_{pe})^2)]\} \tag{11.10}$$

The formula for Γ_β entering Equation (11.9) can be obtained from Equations (11.10) by substituting the subscript α by β. Here, α and β label Stark sublevels of the upper (a) and lower (b) levels involved in the radiative transition, respectively; $\omega_F = 3n_\alpha\hbar F/(2m_e e)$ is the separation between the Stark sublevels caused by the total quasistatic electric field F; the matrix elements of the dipole moment operator **d** are given in Equation (14) from [115]. In Equation (11.10), in the subscripts it was used the notation $\alpha + 1$ and $\alpha - 1$ for the Stark sublevels of the energies $+\hbar\omega_F$ and $-\hbar\omega_F$, respectively (as compared to the energy of the sublevel α).

The quantity γ_p in Equations (11.9) and (11.10) is the sum of the characteristic frequencies of the following processes in plasmas: the electron-ion collision rate γ_{ei} (see, e.g., [134]), the average Landau damping rate γ_L (see, e,g, [135]), and the characteristic frequency γ_{ind} of the nonlinear mechanism of the induced scattering of Langmuir plasmons on ions (see, e.g., [136]):

$$\gamma_p = \gamma_{ei} + \gamma_L + \gamma_{ind} \tag{11.11}$$

The frequency γ_p controls the width of the power spectrum of the Langmuir turbulence.

At the highest density point of KA experiment [109] ($N_e = 1.39 \times 10^{20}$ cm^{-3}, T = 34486 K), the ratio of the thermal-Langmuir-turbulence-caused contribution to the dynamical Stark width of the H$_\alpha$ line to the corresponding contribution by the electron microfield reaches the value ~0.1.

In paper [115], there were calculated Stark profiles of the H$_\alpha$ line with the allowance for the above two effects of the thermal EPT. The formalism of the core generalized theory of the SBHSL [5,7,86] (also allowing for incomplete collisions) was used.

The author of paper [115] modified the formalism of the generalized theory from to allow for the thermal EPT. Namely, he used the distribution of the total quasistatic microfield given by Equation (11.7), thus allowing for the low-frequency thermal EPT, and also added the contribution of the high-frequency thermal EPT to the dynamical broadening operator Φ.

Figure 22 shows the comparison of the experimental FWHM of the H$_\alpha$ line from KA experiment [109] (dots) with the corresponding FWHM yielded by the analytical theory from [115] (solid line). It is seen that the agreement is very good. Even at the highest density point of KA experiment [109] ($N_e = 1.39 \times 10^{20}$ cm^{-3}, T = 34486 K), the theoretical FWHM from [115] differs by just 4.5% from the most probable experimental value and is well within the experimental error margin.

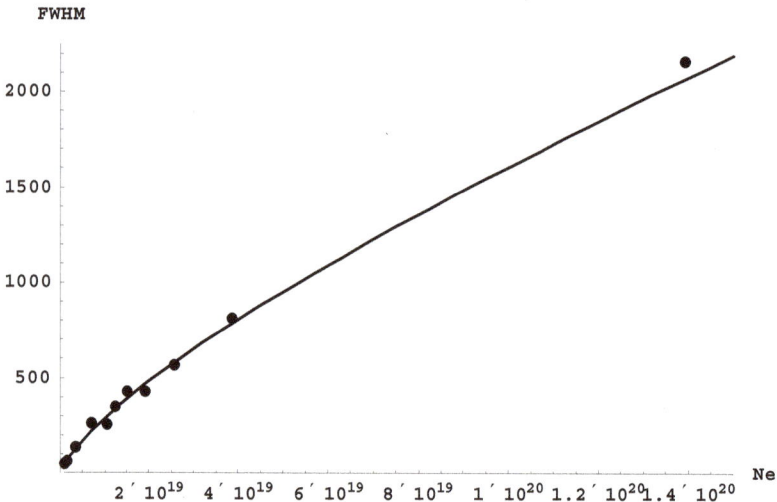

Figure 22. Comparison of the experimental FWHM of the H$_\alpha$ line from Kielkopf–Allard experiment [109] (dots) with the corresponding FWHM yielded by our present analytical theory (solid line). The FWHM is measured in Angstrom, while N_e—in cm^{-3} [115].

11.3. Closing Remarks

Being motivated by the recent benchmark measurements of the FWHM of the H$_\alpha$ line by Kielkopf and Allard [109], which reached the electron densities by two orders of magnitude greater than the corresponding previous benchmark experiments, the author of paper [115] took into account the contribution to the Stark broadening from the *thermal* electrostatic plasma turbulence. It was shown in [115] that this contribution becomes comparable to the corresponding contribution by electron and ion microfields at this range of electron density. As a result, the theoretical FWHM of the H$_\alpha$ line became in a very good agreement with the experimental FWHM of the H$_\alpha$ line by Kielkopf–Allard [109] in the entire range of their electron densities, including the highest electron density $N_e = 1.4 \times 10^{20}$ cm^{-3}.

It was noted in paper [115] that the screening by plasma electrons was taken into account three times. The first time—while separating the collective and individual modes in plasmas: the boundary between them is controlled by the screening by electrons. The second time—while utilizing the APEX distribution for the ion microfield within the modified core generalized theory: the APEX distribution employs plasma ions screened by plasma electrons. The third time—as one of the competing upper cutoffs for the integration over impact parameters of plasma electrons.

The analytical theory from [115] can be also used for calculating Stark profiles and the FWHM of other hydrogen spectral lines. It should be kept in mind though that in the range of $N_e \sim (10^{19} - 2 \times 10^{20})$ cm^{-3}, only three hydrogen spectral lines "survive": H_α, Ly_α, and Ly_β. All of the higher spectral lines of hydrogen merge into a quasicontinuum because of the large Stark broadening at this range of densities. Similarly, in the range of $N_e \sim (2 \times 10^{20} - 10^{22}$ cm$^{-3})$ cm^{-3}, the only one "surviving" spectral line of hydrogen would be Ly_α.

12. The Shape of Spectral Lines of Two-Electron Rydberg Atoms/Ions: A Peculiar Stark Broadening

12.1. Preamble

Effects of the rotating electric field on a hydrogenic atom/ion with the application to the Stark broadening of hydrogen lines in plasmas were studied by various authors. Almost 80 years ago, Spitzer [137–139] found the wave functions of the first excited level ($n = 2$) of a hydrogenic ion in the field produced by a passing ion in the approximation where the passing ion has a rectilinear trajectory (while in reality its trajectory is hyperbolic). In 1967, Ishimura [140] calculated the splitting of the hydrogen Ly-α line in a constant by magnitude electric field rotating with a constant angular velocity. Ishimura considered this kind of the electric field as a model when compared to the real case of the electric field produced by a passing charged particle. The analytical results by Spitzer or by Ishimura, obtained by the direct solution of the corresponding Schrödinger equation, were specific to $n = 2$ and was practically impossible to extend to higher quantum numbers.

In 1970, Lisitsa [141] pointed out the possibility of obtaining the corresponding analytical solution for any principal quantum number n of a hydrogen atom for the case of Ishimura's model electric field, by using the O_4 symmetry of hydrogen atoms that were discovered by Fock [142]. The O_4 symmetry is manifested by the existence of an additional conserved vector quantity: the Runge-Lenz vector [143], also known as Hermann–Bernoulli–Laplace–Runge–Lenz vector [144]. Specifically, Lisitsa showed that this model problem can be reduced to the problem of a hydrogen atom in crossed electric and magnetic fields, whose analytical solution was presented by Demkov, Monozon, and Ostrovsky in 1969 [145]. (Later Lisitsa [8] pointed out that Ishimura's model electric field could be realized for a hydrogen atom in a circularly-polarized laser field.) In 1971, Lisitsa and Sholin [146] used the O_4 symmetry of hydrogen atoms for obtaining the exact analytical solution for the Stark broadening of any hydrogen spectral lines that are caused by one flight of a passing free electron (the binary case). Explicit results were obtained in [146] for the shape of the Ly-alpha line. In 1975, Greene, Cooper, and Smith [147] extended Lisitsa-Sholin's solution to spectral lines of hydrogenic ions under one flight of a passing electron (i.e., also the binary case). In 1996, Derevianko and Oks [148] removed the binary assumption, used by Lisitsa and Sholin, and found the exact analytical solution for the most general, multi-particle description of the interaction of hydrogen atoms with the electron or ion microfield in plasmas. A specific example of the application of the latter formalism was given in [148] for the hydrogen Ly-alpha line. Later, Derevianko and Oks [149] also provided an example of the application of that general formalism to the hydrogen Ly-beta line.

In paper [150] the author used the O_4 symmetry of hydrogenic atoms/ions while considering an isolated two-electron Rydberg atom/ion: namely, the system Z + e + e* (Z being the nuclear charge in atomic units), where the average distance r_{av} of the inner electron e from the nucleus is much smaller than the average distance R_{av} of the outer, Rydberg electron e* from the nucleus. Since the average distance scales with the principal quantum number as n^2, then for the inequality $r_{av} \ll R_{av}$ to be satisfied, it is sufficient that the principal quantum number n_2 of the electron e* would be just several times greater than the principal quantum number n_1 of the electron e. Under the condition $r_{av} \ll R_{av}$, the orbit of the electron e* is very close to the Kepler ellipse of an electron in a hydrogenic atom/ion of the nuclear charge (Z − 1). Therefore, the hydrogenic subsystem Z + e is under the rotating electric field whose magnitude and the angular velocity vary in time.

In paper [150], the results of which we represent here, there was obtained analytically the shape of any spectral line emitted by the subsystem Z + e in this situation. As a particular example, in [150], there was studied in detail the shape of the Ly-alpha line. The obtained analytical expressions yielded, in particular, a *peculiar result*. Namely, as the spectral line splits into several components, the most intense components exhibit a *quadratic* Stark effect with respect to the electric field of the outer electron (though it is linear with respect to the dipole moment of the inner electron), while one would intuitively expect a linear dependence on the electric field of the outer electron since the subsystem Z + e is hydrogenic. In paper [150], there was provided a physical explanation of this peculiar result—the explanation that will be presented below.

12.2. Instantaneous Eigenvalues ("Energies") and Instantaneous Eigenfunctions of the Inner Electron

In paper [150], two-electron Rydberg atoms/ions (i.e., systems Z + e + e*) were treated semiclassically. Namely, there was combined the quantal treatment of the inner electron e with the classical treatment of the outer, Rydberg electron e*.

These atoms/ions emit two sets of spectral lines, such that the frequencies of the spectral lines of one set differ by several orders of magnitude from the frequencies of the spectral lines of the other set. One set of spectral lines is emitted by the outer, Rydberg electron e* at the frequencies equal to $s\Omega$ (s = 1, 2, 3, ...), where Ω is the Kepler frequency of the electron e*. These spectral lines correspond to the radiative transitions between the levels of the principal quantum number n_2 and the principal quantum numbers ($n_2 - s$), where $s \ll n_2$. In the semiclassical approach, analytical expressions for energy levels of the electron e**, including the quantum defect correction, were found, e.g., by Nikitin and Ostrovsky [151], thus allowing for easily obtaining more accurate (than $s\Omega$) results for the radiation frequencies.

This is one of the reasons why in paper [150] the focus was at the spectral lines that were emitted by the inner electron eat the frequencies

$$\omega = [Z^2(1 + 1/m)/2] \, (1/{n_0}^2 - 1/{n_1}^2) \tag{12.1}$$

where m is the mass of the nucleus and n_0 is the principal quantum number of the lower level that is involved in the radiative transition. Here and below, atomic units are used.

Since $n_2 \gg n_1$, then the frequencies of the spectral lines emitted by the inner electron e are by several orders of magnitude greater than the frequencies of the spectral lines emitted by the outer electron e*. Therefore, it should be easier to observe the inner electron spectral lines and especially their structure/splitting—because the frequencies of the lines could be in the optical range—as compared to the outer electron spectral lines, whose frequencies could be, e.g., in the radiofrequency range. This is another reason why in paper [150] the focus was at the shape of spectral lines emitted by the inner electron.

The Hamiltonian of the relative motion of the inner electron e in the Z + e subsystem is as follows

$$H(t) = p^2(1 + 1/m)/2 - Z/r - \mathbf{dF}(t), \mathbf{d} = - (1 + Z/m)\mathbf{r}/(1 + 1/m) \tag{12.2}$$

where $\mathbf{dF}$ is the scalar product (also known as the dot-product) of the dipole moment vector and the vector of the electric field due to the outer electron e*. Here and below, the notation $\mathbf{AB}$ stands for the scalar product of any two vectors $\mathbf{A}$ and $\mathbf{B}$.

The Kepler frequency of the electron e* is

$$\Omega = \{[(Z-1)(1+2/m)]/[R_0{}^3(1+1/m)]\}^{1/2} \tag{12.3}$$

where R_0 is the major semi-axis of the elliptical orbit of the electron e*. At this point, the author of paper [150] proceeded to the reference frame rotating with the time-dependent angular velocity $\Omega_2(t)$ of the electron e* around the direction of the angular momentum **M**. In this rotating frame, the Hamiltonian of the relative motion of the electron e* in the Z + e* subsystem acquires an additional term (see, e.g., book [152]).

$$H(t) = p^2(1+1/m)/2 - Z/r + V(t), V(t) = -\,d\mathbf{F}(t) - \mathbf{M}\Omega_2(t) \tag{12.4}$$

where **M** is the angular momentum of the relative motion of the electron e* in the Z + e* subsystem and $\mathbf{M}\Omega_2$ is its scalar product with the angular velocity vector Ω_2. (In [150] it was noted in passing that in papers [8,141,146], the interaction V(t) erroneously had the opposite sign.) The additional term is equivalent to some of the effective magnetic interaction, so that the Hamiltonian can be rewritten in the form

$$H(t) = p^2(1+1/m)/2 - Z/r - r\mathbf{F}_{eff}(t) - \mu\mathbf{B}_{eff} \tag{12.5}$$

where

$$\mathbf{B}_{eff}(t) = -2c(1+1/m)\,\Omega_2(t)/(1-Z/m^2), \mu = -(1-Z/m^2)\mathbf{M}/[2c(1+1/m)] \tag{12.6}$$

μ being the magnetic dipole moment. In Equation (12.5), the vector $\mathbf{F}_{eff}(t)$ stands for

$$\mathbf{F}_{eff}(t) = -(1+Z/m)\mathbf{F}(t)/(1+1/m), \mathbf{F}(t) = [1 + \varepsilon\cos(\Omega t + \beta)]^2/[(1-\varepsilon^2)^2 R_0{}^2]\mathbf{R}/R \tag{12.7}$$

where ε is the eccentricity of the Kepler ellipse of the electron e* and β is the initial phase.

At any fixed instant of time t, the Hamiltonian in Equations (12.4) or (12.5) corresponds to the Hamiltonian of a hydrogenic atom/ion in crossed electric and magnetic fields, both fields being time-dependent. For the case, where the crossed electric and magnetic fields are time-independent, the classical analytical solution was presented as early as in 1925 in Born's book [153], while the quantum analytical solution for both the energies and the eigenfuctions was obtained by Demkov, Monozon, and Ostrovsky [147] in 1969. Both the classical and quantum solutions that were employed the O_4 symmetry of hydrogenic systems and the Runge-Lenz vector [143]

$$\mathbf{A} = p\times\mathbf{M} - Zr/r \text{ (classically)}, \mathbf{A} = (p\times\mathbf{M} - \mathbf{M}\times p)/2 - Zr/r \text{ (quantally)} \tag{12.8}$$

Lisitsa and Sholin [146], while considering a different problem (a hydrogen atom under one flight of a passing free electron), pointed out that the time-dependent crossed electric and magnetic fields for their problem possess an important property: while, the magnitudes of both fields vary in time, their ratio remains time-independent. The consequence was the existence of the two fixed "directions of quantization". This property allowed Lisitsa and Sholin [146] to reduce the calculation in the rotating frame to just a phase modulation of the atomic oscillator.

While treating the problem considered in paper [150], the author found out a similar situation. Both the electric field **F**(t), imposed by the outer electron e* on the subsystem Z + e, and its angular frequency $\Omega_2(t)$ have the same time dependence, so that the ratio $F(t)/\Omega_2(t)$ remains time-independent. This is a consequence of the fact that $\Omega_2(t) = (1+2/m)\mathbf{M}/[(1+1/m)R^2(t)] = const/R^2(t)$ and $F(t) = 1/R^2(t)$, where R(t) is the absolute value of the radius-vector of the outer, Rydberg electron e*.

In paper [150], there was denoted

$$\Omega_F(t) = (3/2)[r_0/(m_0Z)]^{1/2}\mathbf{F}_{eff}(t), \mathbf{N} = (m_0r_0/Z)^{1/2}\mathbf{A}, r_0 = n^2/(m_0Z) \tag{12.9}$$

where the reduced mass m_0 of the inner electron is

$$m_0 = 1/(1 + 1/m) \qquad (12.10)$$

Physically, the quantity r_0 in Equation (12.9) is the characteristic size of the subsystem $Z + e$; classically it would correspond to the size of the major semi-axis of the elliptical orbit of the *inner* electron e. As for the vector Ω_F, which is perpendicular to the vector Ω_2, physically, its absolute value Ω_F is the instantaneous separation of the Stark sublevels at the instantaneous field-strength $F_{\rm eff}$; classically, Ω_F is the instantaneous frequency of the precession of the Kepler ellipse of the inner electron around the direction of the field $\mathbf{F}_{\rm eff}$.

Following the procedure that is similar to the one from papers [145,146], in paper [150], there were introduced the following four vectors[7]:

$$\mathbf{J}_i(t) = [\mathbf{M} + (-1)^{i+1}\mathbf{N}(t)]/2, \; \boldsymbol{\omega}_i(t) = \Omega_2(t) + (-1)^{i+1}\Omega_F(t); \; i = 1, 2 \qquad (12.11)$$

Then, the interaction in Equation (12.4) can be rewritten in the form:

$$V(t) = -\mathbf{J}_1(t)\boldsymbol{\omega}_1(t) - \mathbf{J}_2(t)\boldsymbol{\omega}_2(t) \qquad (12.12)$$

Since the ratio $q = \Omega_F(t)/\Omega_2(t) = {\rm const}\; F(t)/\Omega_2(t)$ is time-independent, the directions of the vectors $\boldsymbol{\omega}_i(t)$ do not depend on time. Therefore, the Hamiltonian from Equation (12.4) with the interaction $V(t)$ from Equation (12.12) can be diagonalized using the instantaneous wave functions $u_{nn'n''}$, which correspond to a definite instantaneous projection of $\mathbf{J}_1$ on $\boldsymbol{\omega}_1$ (characterized by the quantum number n') and to a definite instantaneous projection of $\mathbf{J}_2$ on $\boldsymbol{\omega}_2$ (characterized by the quantum number n'')—similarly to papers [147,148]. Each of the quantum numbers n' and n'' takes values $-(n-1)/2, -(n-1)/2 + 1, -(n-1)/2 + 2, \dots, (n-1)/2$. The instantaneous eigenvalues ("energies") are as follows

$$E_{nn'n''}(t) = -Z^2/[2n^2(1 + 1/m)] - (n' + n'')\omega(t), \qquad (12.13)$$

The quantity $\omega(t)$ in Equation (12.13) is

$$\omega(t) = |\boldsymbol{\omega}_1(t)| = |\boldsymbol{\omega}_2(t)| = [1 + q^2(n)]^{1/2}\,\Omega_2(t) \qquad (12.14)$$

where

$$q(n) = {\rm const}\; F(t)/\Omega_2(t) = [(1 + 1/m)r_0/Z]^{1/2}\, 3(1 + Z/m)/[2(1 + 2/m)M] = 3n(1 + Z/m)(1 + 1/m)/[2(1 + 2/m)ZM] \qquad (12.15)$$

If one would use the semiclassical expression for the angular momentum of the outer electron $M = L + 1/2$, where L is the corresponding quantum number, then

$$q(n) = 3n(1 + Z/m)(1 + 1/m)/[2(1 + 2/m)Z(L + 1/2)] \qquad (12.16)$$

---

[7]   In paper [150], the intent was to keep the quantum-classical correspondence for the problem of a hydrogenic atom/ion in crossed electric and magnetic fields. There were used some notations from the classical solution presented in the book by Kotkin and Serbo, the 2nd Russian edition, problem 2.37 [154], based partially on the solution of problem 2.36 dealing with the dynamics of the angular momentum **M** and the Runge-Lenz vector **A** in a time-independent electric field [154]. We note that only the 1st Russian edition of this book was published in English [155]; problem 2.37 from [154] was not in the 1st Russian edition and in its English version [155], but problem 2.36 from [154] and its solution is available in [155] numbered as problem 2.32.

It is important to emphasize that

$$q(n) \sim n/[Z(L + 1/2)] \ll 1 \qquad (12.17)$$

as long as

$$L \gg n/Z \qquad (12.18)$$

Under this condition, the directions of the vectors ω_i constitute small angles $\sim q$ with the direction of the vector Ω_2. If it were not for the factor Z, the condition (12.18) would be physically equivalent to the requirement that the outer electron does not penetrate the subsystem $Z + e$, but the factor Z weakens the latter physical requirement. Of course, the semiclassical treatment of the outer, Rydberg electron implies $L \gg 1$.

The instantaneous eigenfunctions in the rotating frame are as follows

$$\chi(t) = u_{nn'n''} \exp\left[-i\omega_0 t + i(n' + n'')[1 + q^2(n)]^{\frac{1}{2}} \int_{t_0}^{t} \Omega_2(s)ds\right], \qquad (12.19)$$

where ω_0 is the unperturbed frequency of the spectral line that is corresponding to the radiative transition of the inner electron between the upper level n_1 and the lower level n_0. Since $\Omega_2(t) = d\theta/dt$, where $\theta(t)$ is the angular variable describing the motion of the outer, Rydberg electron along its elliptical orbit, then Equation (12.18) can be represented in the form:

$$\chi(t) = u_{nn'n''} \exp\{-i\omega_0 t + i(n' + n'')[1 + q^2(n)]^{1/2}[\theta(t) - \theta(t_0)]\} \qquad (12.20)$$

12.3. Spectral Lineshape

The shape $S(\Delta\omega)$ of the corresponding spectral line is given by the Fourier transform of the correlation function $\Phi(\tau)$ of the dipole moments of the subsystem $Z + e$ (see, e.g., book [156]). Here, $\Delta\omega$ is the frequency that is counted from the unperturbed frequency ω_0 of the spectral line. For obtaining analytical results for $S(\Delta\omega)$ in a universal form, below, the author of paper [150] *measured $\Delta\omega$ in units of the Kepler frequency Ω of the outer electron* (given by Equation (12.3)) and *the time variables in units of $1/\Omega$*.

In our case the lineshape can be represented in the form

$$S(\Delta\omega) = \sum_{k=-1,0,1} \sum_{j} |\langle n_1 n_1' n_1' | d_k | n_0 n_0' n_0' \rangle|^2 J_{kj} \qquad (12.21)$$

where j denotes a set of four quantum numbers $(n_1', n_1'', n_0', n_0'')$ and d_k are the spherical components of the dipole moment defined according to Wigner [157] (in particular, $2^{1/2}d_{-1} = d_x + id_y = -2^{1/2}d_{+1}$). The quantity J_k in Equation (12.21) is the following double integral that includes averaging over the initial instant t_0:

$$J_{kj} = [1/(2\pi)^2] \int_{period} d\tau \exp(i\Delta\omega\tau) \int_{period} dt_0 \exp\{-i(\lambda_i + k)[\theta(\tau - t_0) - \theta(-t_0)]\}, \qquad (12.22)$$

where the integrals are taken over any interval of the length 2π (which is the period of the revolution of the outer electron in units of $1/\Omega$) and

$$\lambda_j = (n_0' + n_0'')[1 + q^2(n_0)]^{1/2} - (n_1' + n_1'')[1 + q^2(n_1)]^{1/2} \qquad (12.23)$$

The presence of k (where $k = -1, 0, 1$) in the factor $(\lambda_j + k)$ in Equation (12.22) is due to the transition from the instantaneous eigenfunctions $\chi(t)$ in the rotating frame (given by Equation (20)) to the corresponding eigenfunctions in the laboratory frame. This transition is facilitated by Wigner's D-matrix $D^{(1)}(0, 0, -\theta(t))$—similarly to how this was done in Lisitsa–Sholin's paper [146].

In paper [150], the double integral in Equation (12.22) was reduced to the square of a single integral:

$$J_{kj} = I_{kj}{}^2, I_{ki} = \left[\tfrac{1}{2\pi}\right] \int\limits_{-\pi}^{\pi} d\eta (1 - \varepsilon \cos\eta) \cos\left[\Delta\omega (\eta - \varepsilon \sin\eta) - 2(\lambda_j + k) \tan^{-1}\{[(1+\varepsilon)/(1-\varepsilon)]^{\frac{1}{2}} \tan(\eta/2)\}\right], \qquad (12.24)$$

where ε is the eccentricity of the elliptical orbit of the outer, Rydberg electron (it should be mentioned that the imaginary part of I_{kl}, which was formally present in Equation (12.22), is actually zero). It is worth noting that, semiclassically, the eccentricity of the orbit of the outer electron can be expressed through the corresponding quantum numbers as $\varepsilon = [1 - (L+1/2)^2/n_2{}^2]^{1/2}$.

Thus, in paper [150], there was obtained a universal analytical expression for the shape $S_{kj}(\Delta\omega)$ of the components of the spectral lines, emitted by the inner electron, in terms of the square $J_{kj} = I_{kj}{}^2$ of a single integral $I_{kj}(\Delta\omega)$. The integral $I_{kj}(\Delta\omega)$ depends on two parameters: on the eccentricity ε of the elliptical orbit of the outer electron and on λ_j given by Equation (12.23). The latter depends on the combination of quantum numbers, identifying each component of the spectral line, and on the dimensionless parameters $q(n_1)$ and $q(n_0)$, where $q(n)$ is given by Equation (12.15).

In paper [150], the above general results were illustrated by the example of the Ly-alpha line ($n_1 = 2$, $n_0 = 1$). For this spectral line, Equation (12.23) yields only the following three quantities

$$\lambda = 0, \pm[1 + q^2(2)]^{1/2} \qquad (12.25)$$

The relative intensities of the line components are the same as calculated by Lisitsa and Sholin [146]. They are presented below in Table 5. The sum of relative intensities of all the components is set to unity. The components are identified by their "central frequency" $\lambda_j + k$, which significantly controls the integral in Equation (12.24). It is the frequency, at which the component has the primary maximum of the intensity for $\varepsilon = 0$. We remind that the frequencies are in units of the Kepler frequency Ω of the outer electron (given by Equation (12.3)) and they are counted from the unperturbed frequency of the spectral line.

Table 5. Central frequencies $\lambda_j + k$ and relative intensities of components of the Ly-alpha line emitted by the inner electron. The central frequency is the frequency, at which the component has the primary maximum of the intensity for the zero eccentricity of the orbit of the outer, Rydberg electron. The frequencies are in units of the Kepler frequency Ω of the outer electron (given by Equation (12.3)) and they are counted from the unperturbed frequency of the spectral line. The sum of relative intensities of all the components is set to unity [150].

Central frequency $\lambda_j + k$	$\pm[(1+q^2)^{1/2} - 1]$	$\pm[(1+q^2)^{1/2} + 1]$	± 1	0
Relative intensity	$[(1+q^2)^{1/2} + 1]/[10(1 + q^2)]$	$[(1+q^2)^{1/2} - 1]/[10(1 + q^2)]$	$q^2/[5(1 + q^2)]$	$1/5$

Figure 23 shows the central frequencies of the spectral line components (in units of Ω) versus the parameter q as follows. Components at the central frequencies $\pm[(1 + q^2)^{1/2} - 1]$, $\pm[(1 + q^2)^{1/2} + 1]$, and ± 1 are shown by solid, dashed, and dotted lines, respectively. Of course, the frequency of the 0-component coincides with the abscissa axis.

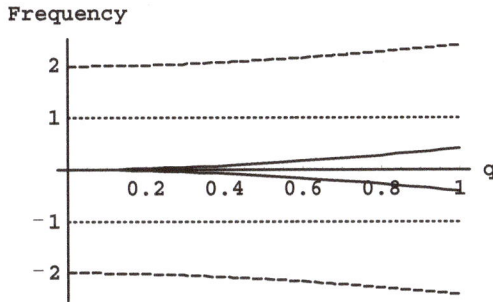

Figure 23. Central frequencies of the Ly-alpha line components versus the parameter q (given by Equations (12.15 and 12.16). Components at the central frequencies $\pm[(1 + q^2)^{1/2} - 1]$, $\pm[(1 + q^2)^{1/2} + 1]$, and ± 1 are shown by solid, dashed, and dotted lines, respectively. The frequency of the 0-component coincides with the abscissa axis. The central frequency is defined as the frequency, at which the component has the primary maximum of the intensity for the zero eccentricity of the orbit of the outer, Rydberg electron. The frequencies are in units of the Kepler frequency Ω of the outer electron (given by Equation (12.3)) and they are counted from the unperturbed frequency of the spectral line [150].

Figure 24 presents the relative intensities of the Ly-alpha line components versus the parameter q, as follows. Intensities of the components having the central frequencies $\pm[(1 + q^2)^{1/2} - 1]$, $\pm[(1 + q^2)^{1/2} + 1]$, ±1, and 0 are shown by solid, dashed, dotted, and dash-dotted lines, respectively.

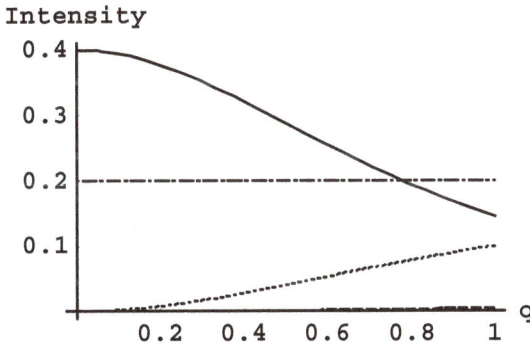

Figure 24. Relative intensities of the Ly-alpha line components versus the parameter q (given by Equations (12.15) and (12.16). Intensities of the components having the central frequencies $\pm [(1 + q^2)^{1/2} - 1]$, $\pm [(1 + q^2)^{1/2} + 1]$, ± 1, and 0 are shown by solid, dashed, dotted, and dash-dotted lines, respectively. The sum of relative intensities of all components is set to unity [150].

It is seen that at $q^2 \ll 1$, about 80% the entire intensity of the spectral line is represented by the components at the central frequencies $\pm[(1 + q^2)^{1/2} - 1]$. At $q^2 \ll 1$, the central frequency is approximately equal to $q^2/2$. Since $q = \Omega_F(t)/\Omega_2(t) = \text{const } F(t)/\Omega_2(t)$, the central frequency is proportional to the square $F^2(t)$ of the instantaneous electric field that is produced by the outer, Rydberg electron. In other words, at $q^2 \ll 1$—which is the actual case as long as $L^2 \gg (n/Z)^2$, according to Equation (12.18)—the primary result of the electric field produced by the outer electron is the *quadratic* Stark effect, which is *counter-intuitive*. Indeed, intuitively one would think that since the subsystem Z + e is hydrogenic, the Stark effect in an external electric field should be linear.

This small quadratic shift of the maximum of the profile of the component labeled as $[(1 + q^2)^{1/2} - 1]$, which is the primary blue component, as illustrated in Figure 25. It shows a magnified part of the profile (Sbprimary) of this component, calculated, as an example, at $\varepsilon = 0$ and $q^2 = 1/10$, close to the unperturbed frequency.

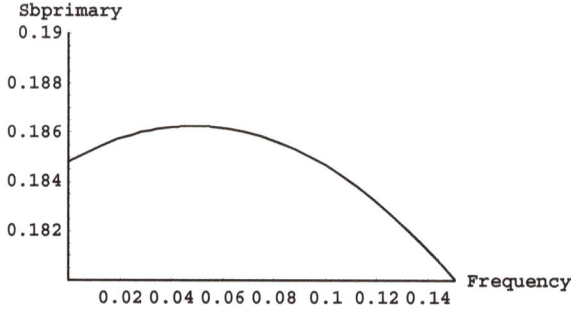

Figure 25. Magnified part of the primary blue component of the Ly-alpha line calculated at $\varepsilon = 0$ and $q^2 = 1/10$. The frequencies are in units of the Kepler frequency Ω of the outer electron (given by Equation (12.3)) [150].

The physical explanation of this counter-intuitive result is that at $L \gg n/Z$, the dominant effect is the rotation of the electric field that is produced by the outer electron. As a consequence, the primary effect in the splitting of the spectral line is due to the *amplitude modulation* of the (Z+e)-oscillator by the rotating electric field (i.e., the change in the amplitude of the emitted electromagnetic wave) and it results in the quadratic Stark effect. The other two well-known contributions of electrons to the lineshape, described, e.g., in review [8]—the phase modulation (caused by the time variation of the absolute value of the electric field) and the nonadiabatic effects (causing the virtual transitions between the Stark sublevels)—are secondary in this situation.

The shape of the entire Ly-alpha line is as follows:

$$J(\Delta\omega, \varepsilon) = \{[(1 + q^2)^{1/2} + 1]/[10(1 + q^2)]\} \{I^2[\Delta\omega, \varepsilon,$$
$$(1 + q^2)^{1/2} - 1] + I^2[\Delta\omega, \varepsilon, 1 - (1+q^2)^{1/2}]\} + \{[(1 + q^2)^{1/2} - 1]/[10(1 + q^2)]\} \{I^2[\Delta\omega, \quad (12.26)$$
$$\varepsilon, (1 + q^2)^{1/2} + 1] + I^2[\Delta\omega, \varepsilon, - (1+q^2)^{1/2} - 1]\} + \{q^2/[5(1 + q^2)]\} \{I^2[\Delta\omega, \varepsilon, 1] + I^2[\Delta\omega, \varepsilon, - 1]\} + (1/5) \delta(\Delta\omega)$$

where the integrals $I_{jk}(\Delta\omega, \varepsilon, \lambda_j + k)$ are given in Equation (12.24). In the last term in Equation (12.26), $\delta(\Delta\omega)$ is the Dirac δ-function, reflecting the fact that the rotating electric field that is produced by the outer electron does not broaden the "unshifted" component of the Ly-alpha line. Physically, this "unshifted" component has the natural width, i.e., the component has a relatively narrow Lorentzian profile whose width is the inverse of the natural lifetime of the $n = 2$ state of the Z+e subsystem.

Figure 26 presents, as an example, the total profile (Stot) of all "shifted" components at $\varepsilon = 0$ and two values of q^2: $q^2 = 1/5$ (solid line) and $q^2 = 1/25$ (dashed line). It is seen that at $q^2 = 1/5$, the profile exhibits a pedestal (or "shoulders") at $|\Delta\omega| = (1 - 1.4)$, while at $q^2 = 1/25$, secondary maxima show up more distinctly.

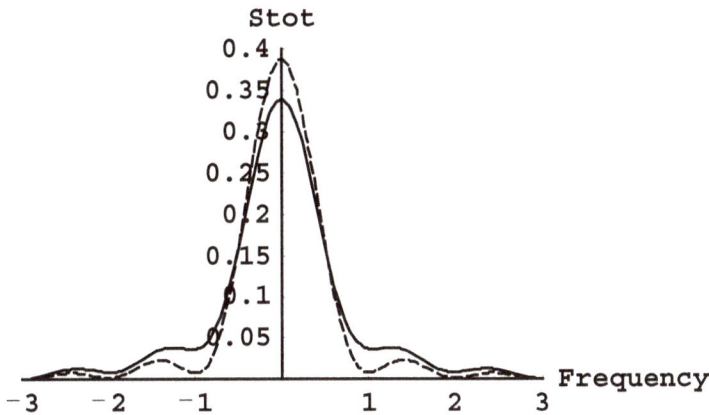

Figure 26. Total profile of all "shifted" components at $\varepsilon = 0$ and two values of q^2: $q^2 = 1/5$ (solid line) and $q^2 = 1/25$ (dashed line). The frequencies are in units of the Kepler frequency Ω of the outer electron (given by Equation (12.3)) [150].

Figure 27 shows, as an example, the total profile (Stot) of all "shifted" components at $q^2 = 1/5$ and two values of ε: $\varepsilon = 0.9$ (solid line) and $\varepsilon = 0$ (dashed line). It is seen that a relatively large eccentricity ($\varepsilon = 0.9$) of the orbit of the outer electron makes the secondary maxima more pronounced when compared to the circular orbit.

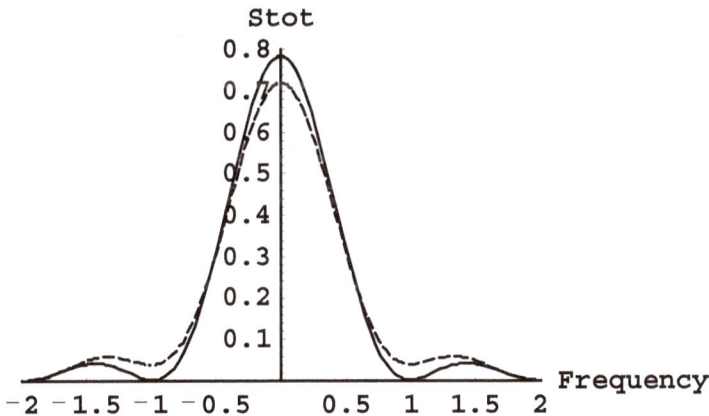

Figure 27. Total profile of all "shifted" components at $q^2 = 1/5$ and two values of ε: $\varepsilon = 0.9$ (solid line) and $\varepsilon = 0$ (dashed line). The frequencies are in units of the Kepler frequency Ω of the outer electron (given by Equation (12.3)) [150].

Finally, in paper [150], the above dynamical Stark broadening (of the spectral lines of the subsystem Z + e) by the Rydberg electron (for brevity, DSBRE) was compared with competing broadening mechanisms/models both for purely atomic (no plasma) experiments and for some important astrophysical and laboratory plasmas. As an example, there was used the Ly-alpha line of the ionized helium (He II 30.38 nm) that was emitted by the Z + e subsystem that experiences the DSBRE, the outer electron having the principal quantum number n = 10. (The ratio of the average distance of the outer electron from the nucleus to the average distance of the inner electron from the nucleus is $(10/2)^2 = 25 \gg 1$.)

The width due to the DSBRE $\Delta\omega_{DSBRE}$ is approximately equal to the Keppler frequency of the outer electron:

$$\Delta\omega_{DSBRE} \sim \omega_{au}/n^3 = 4.13 \times 10^{16} \text{ s}^{-1}/10^3 = 4.13 \times 10^{13} \text{ s}^{-1} \tag{12.27}$$

where ω_{au} is the atomic unit of frequency.

The natural width is $\Delta\omega_{nat} = 1.25 \times 10^9$ s^{-1}, so that the DSBRE $\Delta\omega_{DSBRE}$ exceeds the natural width by four orders of magnitude.

The Doppler width for purely atomic (no plasma) experiments at the room temperature, i.e., T = 0.026 eV, is $\Delta\omega_D = 0.958 \times 10^{11}$ s^{-1}.

The Doppler width for the conditions of He II emitting regions of solar flares (chromosphere and transition region plasmas), where T~2 eV, is $\Delta\omega_D = 8.42 \times 10^{11}$ s^{-1}.

The Doppler width for the conditions of edge plasmas of magnetic fusion devices, where T~4 eV, is $\Delta\omega_D = 1.19 \times 10^{12}$ s^{-1}.

Thus, in all of the above three situations, the DSBRE $\Delta\omega_{DSBRE}$ exceeds the Doppler broadening by one or two orders of magnitude.

Then, in paper [150], the DSBRE was compared with the dynamical Stark broadening by plasma electrons (DSBE) and with the dynamical Stark broadening by plasma ions (DSBI) – for the conditions of He II emitting regions of solar flares, where the electron density is ~10^{13} cm^{-3} and T = 2 eV, and for the conditions of edge plasmas of magnetic fusion devices, where the electron density is ~10^{14} cm^{-3} and T = 4 eV.

For the conditions of He II emitting regions of solar flares, the DSBE width is $\Delta\omega_{DSBE} = 1.2 \times 10^8$ s^{-1} and the DSBI is $\Delta\omega_{DSBI} = 1.9 \times 10^9$ s^{-1}.

For the conditions of the edge plasmas of magnetic fusion devices, the DSBE width is $\Delta\omega_{DSBE} = 4.6 \times 10^8$ s^{-1} and the DSBI is $\Delta\omega_{DSBI} = 1.3 \times 10^{10}$ s^{-1}.

Thus, in both of the above important astrophysical and laboratory plasmas situations the DSBRE $\Delta\omega_{DSBRE}$ exceeds the DSBE and the DSBI by three or four or five orders of magnitude.

It is important to emphasize that the DSBRE and the DSBE/DSBI are *different kinds of the Stark broadening*, and that the DSBRE was never included in the DSBE/DSBI.

12.4. Closing Remarks

In paper [150], there was considered an isolated two-electron Rydberg atom/ion, namely the system Z + e + e*, where the average distance r_{av} of the inner electron e from the nucleus is much smaller than the average distance R_{av} of the outer, Rydberg electron e* from the nucleus. There was analytically calculated the shape of spectral lines that are emitted by the inner electron under the rotating electric field of the outer electron—by using the O_4 symmetry of hydrogenic atoms/ions.

The obtained analytical expressions for the shape of the Ly-alpha line, which was used as an example for the detailed study, yielded, in particular, a *peculiar result*. Namely, as the spectral line splits into several components, the most intense components exhibit a *quadratic* Stark effect with respect to the electric field of the outer electron (though it is linear with respect to the dipole moment of the inner electron), while one would intuitively expect a linear dependence of the electric field of the outer electron, since the subsystem Z + e is hydrogenic. The physical reason is that the rotation of the electric field that is produced by the outer electron causes the predominant effect to be the amplitude modulation of the (Z + e)-oscillator, and it results in the quadratic Stark effect.

The obtained spectral line shapes differ from the line shapes due to the dynamical Stark broadening in plasmas (by plasma electrons and by plasma ions). The dynamical Stark broadening in plasmas can lead to a variety of line shapes: from the simplest Lorentzian (corresponding to the simplest effect where the upper state of the radiating atom/ion exponentially falls in time getting "killed" by collisions with plasma electrons) to much more complicated shapes (corresponding to the combinations of the phase modulation, the amplitude modulation, and the nonadiabatic effects)—see, e.g., the exact analytical results from papers [146–149] and the review in book [5]. From the physical point of view, the primary distinction between the DSBP and the broadening that is calculated in the paper [150] is that in the former the motion of perturbing plasma electrons is aperiodic, while in the latter the motion of the perturbing outer electron is periodic or quasi-periodic. This physical difference translates into the difference in line shapes. It should be noted that the spectral line broadening that is calculated in the paper [150] is generally much greater than the natural broadening, so that the latter has been neglected.

A final remark: the influence of the inner electron on the outer electron translates into the allowance for the fact that the outer electron moves in a dipole potential of the subsystem "nucleus + inner electron". Studies of this effect (see, e.g., paper [103] and references therein) showed that the result is a relatively small correction to the motion of the outer electron. In the present paper, the focus is on how the motion of the outer electron modulates the motion of the inner electron. The modulation is a correction to the unperturbed motion of the inner electron. In paper [150] this correction (the modulation) has been calculated for the elliptical trajectory of the outer electron. The allowance for dipole potential of the subsystem "nucleus + inner electron" leads to relatively small changes of the elliptical trajectory of the outer electron. The allowance of the effect of such changes on the modulation of the motion of the inner electron would constitute a correction to the correction. Therefore, it was legitimate to consider in paper [150] the first non-vanishing effect of the modulation and to disregard the higher order corrections.

13. Conclusions

The analytical results that were reviewed in this paper can be best summarized by subdividing them in the following groups.

13.1. Fundamental Results for Atomic Physics

It was shown that, for hydrogen/deuterium atoms in a nonuniform electric field, the center-of-mass motion, generally speaking, could not be separated from the relative motion. However, it was also shown that these two motions can be separated by using the approximate analytical method of separating rapid and slow subsystems. This has also practical applications noted in Section 13.3 below.

Also it was demonstrated that for isolated two-electron Rydberg atoms/ion, the shape of spectral lines exhibits a *peculiar Stark broadening*. Namely, as the spectral line splits into several components, the most intense components exhibit a quadratic Stark effect with respect to the electric field of the outer electron (though it is linear with respect to the dipole moment of the inner electron), while one would intuitively expect a linear dependence on the electric field of the outer electron, since the subsystem Z + e is hydrogenic.

13.2. Fundamental Results for Plasma Spectroscopy

It was shown that for the Stark broadening of hydrogenic spectral lines by plasma electrons, the allowance for the more realistic, non-rectilinear trajectories of plasma electrons (caused by their interaction with the dipole moment of the radiating atom) becomes significant already at $N_e \sim 10^{17}$ cm^{-3}, and very significant at higher densities.

Also, it was demonstrated that *the allowance for penetrating ions* (i.e., the ions inside the bound electron cloud of the radiating atom) causes *an additional red shift* of hydrogenic spectral lines spectral lines. *This shift can be greater that even the sum of all standard, previously known shifts*, and can eliminate a huge discrepancy (by factor between two and five) between the previous theories and experiments.

Further, it was shown that for the Stark broadening of hydrogen lines in plasmas of electron densities up to or more than $N_e \sim 10^{20}$ cm^{-3}, *the electrostatic plasma turbulence at the thermal level makes a significant contribution the Stark width*. Previously, there were no theories of the Stark broadening of hydrogen lines in plasmas of electron densities $N_e \sim 10^{19}$ cm^{-3} and higher. The allowance for the electrostatic plasma turbulence at the thermal level is very important for this kind of electron densities and it leads to a very good agreement with the benchmark experiment.

13.3. Practical Applications

It was shown that magnetic-field-caused modifications of trajectories of plasma electrons (i.e., the allowance for the fact that the trajectory becomes helical) lead to a quite dramatic change of the ratio of the intensities of the π-component to the σ-component of the Zeeman triplet, and also to some additional shift of these components. In laboratory plasmas, it is important for magnetic fusion. In astrophysical plasmas, it is important for white dwarfs.

In addition, it was demonstrated that the Stark width of hydrogen/deuterium spectral lines, calculated with the allowance for helical trajectories of perturbing electrons does not depend on the magnetic field for the case of sufficiently strong magnetic fields. Further, it was shown that, depending on the particular spectral line and on plasma parameters, the Stark width, as calculated with the allowance for helical trajectories of perturbing electrons, can be either by orders of magnitude smaller, or of the same order, or several times higher than the corresponding Stark width, calculated for rectilinear trajectories of perturbing electrons. Such a variety of outcomes is a counterintuitive result and it should motivate astrophysicists for a very significant revision of all the existing calculations of the broadening of hydrogen lines in DA white dwarfs.

Also there were calculated analytically *Lorentz–Doppler profiles* of hydrogen/deuterium spectral lines for any angle of observation (with respect to the magnetic field) and any magnetic field strength. *Two counterintuitive results* have been obtained. First, at the perpendicular or close to the perpendicular direction of observation, *the π-components* of the hydrogen/deuterium spectral lines *get suppressed compared to the σ-components*. Second, *the width* of the Lorentz–Doppler profiles turned out to be *a non-monotonic function of the magnetic field* for observations that are perpendicular to **B**. In laboratory plasmas, these results are important for *magnetic fusion*. In astrophysics, these results are important for *solar plasmas*.

Further, there was developed an advanced analytical theory of the Stark broadening of hydrogen/deuterium spectral lines by a Relativistic Electron Beam (REB). It was demonstrated that, especially sensitive to the final stage of the development of the REB, would be the ratio of widths of σ- and π-components, which could be determined by the polarization analysis and to serve as the detection tool. It was also shown that the early stage of the development of the REB could be detected by observing the formation of the L-dips in spectral line profiles. The observation of the L-dips, which manifest the development of strong Langmuir waves caused by the REB, could be an important tool for the early detection and the mitigation of the problem of REB in magnetic fusion machines, such as, e.g., tokamaks.

It was also demonstrated that, *in strongly-magnetized plasmas, the Inglis–Teller diagnostic method* (based on the principal quantum number of the last observed line in the spectral series of hydrogen/deuterium lines) *yields the quantity* B^2T, where B is the magnetic field and T is the atomic temperature. This is in distinction to non- or weakly-magnetized plasmas, where this method yielded the electron temperature. In laboratory plasmas, this result is important for *magnetic fusion*. In astrophysics this result is important for *solar plasmas*.

Further, *the method for measuring the electron density based on the asymmetry* of hydrogenic spectral lines in dense plasmas was revised by *allowing for the contribution on the penetrating ions* to the asymmetry. It was shown that, without this allowance, there *would be significant errors* in deducing the electron density by this methods *in plasmas of* $N_e \sim 10^{18}$ *cm*$^{-3}$ *and higher*.

Finally, it should be noted that the fundamental result of *the approximate analytical separation of the center-of-mass and relative motions* in hydrogen/deuterium atoms in a non-uniform electric field, presented in detail in Section 3 and mentioned in Section 13.1 of conclusions, also has *practical applications to the ion dynamical Stark broadening* of hydrogen/deuterium lines in plasmas. In laboratory plasmas, it is important for *magnetic fusion* plasmas and *radiofrequency discharges*. In astrophysical plasmas, it is important for atmospheres of *flare stars*.

Conflicts of Interest: The author declares no conflict of interest.

References

1. Oks, E. *Plasma Spectroscopy: The Influence of Microwave and Laser Fields*; Series on Atoms and Plasmas; Springer: New York, NY, USA, 1995; Volume 9.
2. Griem, H.R. *Principles of Plasma Spectroscopy*; Cambridge University Press: Cambridge, UK, 1997.
3. Salzman, D. *Atomic Physics in Hot Plasmas*; Oxford University Press: Oxford, UK, 1998.
4. Fujimoto, T. *Plasma Spectroscopy*; Clarendon Press: Oxford, UK, 2004.
5. Oks, E. *Stark Broadening of Hydrogen and Hydrogenlike Spectral Lines in Plasmas: The Physical Insight*; Alpha Science International: Oxford, UK, 2006.
6. Kunze, H.-J. *Introduction to Plasma Spectroscopy*; Springer: Berlin, Germany, 2009.
7. Oks, E. *Diagnostics of Laboratory and Astrophysical Plasmas Using Spectral Lines of One-, Two-, and Three-Electron Systems*; World Scientific: Hackensack, NJ, USA, 2017.
8. Lisitsa, V.S. Stark broadening of hydrogen lines in plasmas. *Sov. Phys. Uspekhi* **1977**, *122*, 603. [CrossRef]
9. Post, D.E.; Votta, L.G. Computational science demand a new paradigm. *Phys. Today* **2005**, *58*, 35–41. [CrossRef]
10. Oks, E. Theories, experiments, and simulations of spectral line shapes: Pitfalls in the network. *AIP Conf. Proc.* **2010**, *1290*, 6–13.
11. Griem, H.R.; Shen, K.Y. Stark broadening of hydrogenic ion lines in a plasma. *Phys. Rev.* **1961**, *122*, 1490–1496. [CrossRef]
12. Griem, H.R. *Plasma Spectroscopy*; McGraw-Hill: New York, NY, USA, 1964.
13. Griem, H.R. *Spectral Line Broadening by Plasmas*; Academic Press: Cambridge, MA, USA, 1974; Chapter II.3b.
14. Sanders, P.; Oks, E. Allowance for more realistic trajectories of plasma electrons in the Stark broadening of hydrogenlike spectral lines. *J. Phys. Commun.* **2018**, *2*, 035033. [CrossRef]
15. Galitski, V.; Karnakov, B.; Kogan, V.; Galitski, V., Jr. *Exploring Quantum Mechanics*; Oxford University Press: Oxford, UK, 2013.
16. Kotkin, G.L.; Serbo, V.G. *Collection of Problems in Classical Mechanics*; 1971; Pergamon: Oxford, UK, problem 2.3.
17. Oks, E. Center-of-mass effects for hydrogen atoms in a nonuniform electric field: Applications to magnetic fusion, radiofrequency discharges, and flare stars. *J. Phys. Commun.* **2018**, *2*, 045005. [CrossRef]
18. Kotkin, G.L.; Serbo, V.G. *Collection of Problems in Classical Mechanics*; Pergamon: Oxford, UK, 1971; problem 2.22.
19. Fox, K. Classical motion of an electron in an electric-dipole field II. Point dipole case. *J. Phys. A* **1968**, *1*, 124–127. [CrossRef]
20. Abramov, V.A.; Lisitsa, V.S. *Sov. J. Plasma Phys.* **1977**, *3*, 451.
21. Seidel, J. Hydrogen stark broadening by ion impacts on moving emitters. *Z. Naturforsch.* **1979**, *34*, 1385–1397. [CrossRef]
22. Stehle, C.; Feautrier, N. Stark broadening of the Hα line of hydrogen at low densities: Quantal and semiclassical results. *J. Phys. B* **1984**, *17*, 1477–1490. [CrossRef]
23. Derevianko, A.; Oks, E. Generalized theory of ion impact broadening in magnetized plasmas and its applications for tokamaks. *Phys. Rev. Lett.* **1994**, *73*, 2059–2062. [CrossRef] [PubMed]
24. Derevianko, A.; Oks, E. Ion impacts on moving emitters: A convergent theory of anisotropic broadening in high-temperature plasmas. *J. Quant. Spectr. Radiat. Transf.* **1995**, *54*, 137–142. [CrossRef]
25. Derevianko, A.; Oks, E. Dual purpose diagnostics of edge plasmas of tokamaks based on a novel spectroscopic effect. *Rev. Sci. Instrum.* **1997**, *68*, 998–1001. [CrossRef]
26. Kolb, A.C.; Griem, H.R. Theory of line broadening in multiplet spectra. *Phys. Rev.* **1958**, *111*, 514–521. [CrossRef]
27. Baranger, M. Problem of overlapping lines in the theory of pressure broadening. *Phys. Rev.* **1958**, *111*, 494–504. [CrossRef]
28. Kepple, P.; Griem, H.R. Improved stark profile calculations for the hydrogen lines Hα, Hβ, Hγ, and Hδ. *Phys. Rev.* **1968**, *173*, 317–325. [CrossRef]
29. Griem, H.R. *Spectral Line Broadening by Plasmas*; Academic Press: Cambridge, MA, USA, 1974; Chapter II.3a.
30. Pospieszczyk, A. Spectroscopic diagnostics of tokamak edge plasmas. *Phys. Scr.* **2005**, *2005*, 71–82. [CrossRef]
31. Gershberg, R.E. *Solar-Type Activity in Main-Sequence Stars*; Springer: Berlin, Germany, 2005.

32. Oks, E.; Gershberg, R.E. Flare stars—A favorable object for studying mechanisms of nonthermal astrophysical phenomena. *Astrophys. J.* **2016**, *819*, 16. [CrossRef]

33. Bengston, R.D.; Tannich, J.D.; Kepple, P. Comparison between measured and theoretical stark-broadened profiles of H_6–H_{12} emitted from a low-density plasma. *Phys. Rev. A* **1970**, *1*, 532–533.

34. Bengtson, R.D.; Chester, G.R. Observation of shifts of hydrogen lines. *Astrophys. J.* **1972**, *178*, 565–569. [CrossRef]

35. Himmel, G. Plasma effects in the spectrum of high Balmer lines. *J. Quant. Spectrosc. RadIAT. Transf.* **1976**, *16*, 529–536. [CrossRef]

36. Nussbaumer, H.; Bieri, L. *Discovering the Expanding Universe*; Cambridge University Press: Cambridge, UK, 2009.

37. Parigger, C.G.; Plemmons, D.H.; Oks, E. Balmer series H-beta measurements in a laser-induced hydrogen plasma. *Appl. Opt.* **2003**, *42*, 5992–6000. [CrossRef] [PubMed]

38. Oks, E. New source of the red shift of highly-excited hydrogenic spectral lines in astrophysical and laboratory plasmas. *J. Astrophys. Aerosp. Technol.* **2017**, *5*. [CrossRef]

39. Sanders, P.; Oks, E. Estimate of the Stark shift by penetrating ions within the nearest perturber approximation for hydrogenlike spectral lines in plasmas. *J. Phys. B* **2017**, *50*, 245002. [CrossRef]

40. Pittman, T.L.; Fleurier, C. Plasma shifts of the He II H_α and P_α lines. *Phys. Rev. A* **1986**, *33*, 1291–1296. [CrossRef]

41. Griem, H.R. Shift of hydrogen lines from electron collisions in dense plasmas. *Phys. Rev. A* **1983**, *28*, 1596–1601. [CrossRef]

42. Boercker, D.B.; Iglesias, C.A. Static and dynamic shifts of spectral lines. *Phys. Rev. A* **1984**, *30*, 2771–2774. [CrossRef]

43. Griem, H.R. Shift of hydrogen and ionized-helium lines from $\Delta n = 0$ interactions with electrons in dense plasmas. *Phys. Rev. A* **1988**, *38*, 2943–2952. [CrossRef]

44. Renner, O.; Salzmann, D.; Sondhauss, P.; Djaoui, A.; Krousky, E.; Förster, E. Experimental evidence for plasma shifts in Lyman series of aluminum. *J. Phys. B* **1998**, *31*, 1379–1390. [CrossRef]

45. Berg, H.F.; Ali, A.W.; Lincke, R.; Griem, H.R. Measurements of Stark profiles of neutral and ionized helium and hydrogen lines from shock-heated plasmas in electromagnetic T tubes. *Phys. Rev.* **1962**, *125*, 199–206. [CrossRef]

46. Oks, E. New type of shift of hydrogen- and hydrogenlike spectral lines. *J. Quant. Spectrosc. Radiat. Transf.* **1997**, *58*, 821–826. [CrossRef]

47. Könies, A.; Günter, S. Asymmetry and shifts of the Lα- and Lβ-lines of hydrogen. *J. Quant. Spectrosc. Radiat. Transf.* **1994**, *52*, 825–830. [CrossRef]

48. Günter, S.; Könies, A. Shifts and asymmetry parameters of hydrogen Balmer lines on dense plasmas. *Phys. Rev. E* **1977**, *55*, 907–911. [CrossRef]

49. Demura, A.V.; Helbig, V.; Nikolic, D. *Spectral Line Shapes, 16th ICSLS*; Back, C.A., Ed.; AIP Conference Proceedings; American Institute of Physics: New York, NY, USA, 2002; Volume 645.

50. Sholin, G.V. On the nature of the asymmetry of the spectral line profiles of hydrogen in dense plasmas. *Opt. Spectosc.* **1969**, *26*, 275–282.

51. Komarov, I.V.; Ponomarev, L.I.; Yu, S. *Slavyanov, Spheroidal and Coulomb Spheroidal Functions*; Nauka: Moscow, Russia, 1976.

52. Held, B. Electric microfield distribution in multicomponent plasmas. *J. Phys.* **1984**, *45*, 1731–1750. [CrossRef]

53. Held, B.; Deutsch, C.; Gombert, M.-M. Low-frequency electric microfield in dense and hot multicomponent plasmas. *Phys. Rev. A* **1984**, *29*, 880–895. [CrossRef]

54. Podder, N.K.; Clothiaux, E.J.; Oks, E. A method for density measurements employing an asymmetry of lineshapes in dense plasmas and its implementation in a vacuum spark discharge. *J. Quant. Spectrosc. Radiat. Transf.* **2000**, *65*, 441–453. [CrossRef]

55. Parigger, C.G.; Swafford, L.D.; Woods, A.C.; Surmick, D.M.; Witte, M.J. Asymmetric hydrogen beta electron density diagnostics of laser-induced plasma. *Spectrochim. Acta Part B* **2014**, *99*, 28–33. [CrossRef]

56. Sanders, P.; Oks, E. Improving the method of measuring the electron density via the asymmetry of hydrogenic spectral lines in plasmas by allowing for penetrating ions. *Atoms* **2018**, *6*, 21. [CrossRef]

57. Oks, E.; Uzer, T. A robust perturbation theory for degenerate states based on exact constants of the motion. *Europhys. Lett.* **2000**, *49*, 554–557. [CrossRef]

58. Kryukov, N.; Oks, E. Super-generalized Runge-Lenz vector in the problem of two Coulomb or Newton centers. *Phys. Rev. A* **2012**, *85*. [CrossRef]

59. Oks, E. Role of Lorentz-Stark broadening of hydrogen spectral lines in magnetized plasmas: Applications to magnetic fusion and solar physics. *J. Quant. Spectr. Radiat. Transf.* **2015**, *156*, 24–35. [CrossRef]

60. Welch, B.L.; Griem, H.R.; Terry, J.; Kurz, C.; LaBombard, B.; Lipschultz, B.; Marmar, E.; McCracken, J. Density measurements in the edge, divertor, and X-point regions of Alcator C-Mod from Balmer series emission. *Phys. Plasmas* **1995**, *2*, 4246–4251. [CrossRef]

61. Brooks, N.H.; Lisgo, S.; Oks, E.; Volodko, D.; Groth, M.; Leonard, A.W.; DIII-D Team. Benchmarking of alternate theories for Stark broadening against experimental data from DIII-D diagnostics. *Plasma Phys. Rep.* **2009**, *35*, 112–117. [CrossRef]

62. Oks, E. *Atomic Processes in Basic and Applied Physics*; Shevelko, V., Tawara, H., Eds.; Springer: Heidelberg, Germany, 2012; Chapter 15.

63. Oks, E.; Bengtson, R.D.; Touma, J. Application of the generalized theory of Stark broadening to experimental highly-excited Balmer lines from a radio-frequency discharge. *Contrib. Plasma Phys.* **2000**, *40*, 158–161. [CrossRef]

64. Feldman, U.; Doschek, G.A. The emission spectrum of the hydrogen Balmer series observed above the solar limb from SKYLAB. II. Active regions. *Astrophys. J.* **1977**, *212*, 913–922. [CrossRef]

65. Galushkin, Y.I. Electrodynamic broadening of the spectral lines with the linear Stark effect. *Sov. Astron.* **1970**, *14*, 301–309.

66. Sanders, P.; Oks, E. Lorentz–Doppler profiles of hydrogen/deuterium lines for magnetic fusion: Analytical solution for any angle of observation and any magnetic field strength. *J. Phys. Commun.* **2017**, *1*, 055011. [CrossRef]

67. Neu, R.; Summers, H.P.; Ralchenko, Y. Spectroscopic diagnostics of magnetic fusion plasmas. *J. Phys. B* **2010**, *43*, 140201. [CrossRef]

68. Rosenberg, F.D.; Feldman, U.; Doschek, G.A. The emission spectrum of the hydrogen Balmer series observed above the solar limb from SKYLAB. I. A quiet Sun and a polar coronal hole. *Astrophys. J.* **1977**, *212*, 905–912. [CrossRef]

69. Inglis, D.R.; Teller, E. Ionic depression of series limits in one-electron spectra. *Astrophys. J.* **1939**, *90*, 439–448. [CrossRef]

70. Hey, J. Does atomic polarizability play a role in hydrogen radiorecombination spectra from Galactic H II regions? *J. Phys. B* **2013**, *46*, 175702. [CrossRef]

71. Oks, E. Aspects of Lorentz-Stark broadening of hydrogen spectral lines in magnetized, turbulent and non-turbulent plasmas important for magnetic fusion and solar physics Intern. *Rev. Atom. Mol. Phys.* **2013**, *4*, 105–119.

72. Rosato, J.; Marandet, Y.; Stamm, R. Stark broadening by Lorentz fields in magnetically confined plasmas. *J. Phys. B* **2014**, *47*, 105702. [CrossRef]

73. Guenot, D.; Gustas, D.; Vernier, A.; Beaurepaire, B.; Böhle, F.; Bocoum, M.; Losano, M.; Jullien, A.; Lopez-Martins, A.; Lifschitz, A.; et al. Relativistic electron beams driven by kHzsingle cycle light pulses. *Nat. Photonics* **2017**, *11*, 293–296. [CrossRef]

74. Kurkin, S.A.; Hramov, A.E.; Koronovskii, A.A. Microwave radiation power of relativistic electron beam with virtual cathode in the external magnetic field. *Appl. Phys. Lett.* **2013**, *103*, 043507. [CrossRef]

75. De Jagher, P.C.; Sluijter, F.W.; Hopman, H.J. Relativistic electron beams and beam-plasma interactions. *Phys. Rep.* **1988**, *167*, 177–239. [CrossRef]

76. Boozer, A.H. Runaway electrons and ITER. *Nucl. Fusion* **2017**, *57*, 056018. [CrossRef]

77. Decker, J.; Hirvijoki, E.; Embreus, O.; Peysson, Y.; Stahl, A.; Pusztai, I.; Fülöp, T. Numerical characterization of bump formation in the runaway electron tail. *Plasma Phys. Control. Fusion* **2016**, *58*, 025016. [CrossRef]

78. Smith, H.; Helander, P.; Eriksson, L.-G.; Anderson, D.; Lisak, M.; Andersson, F. Runaway electrons and the evolution of the plasma current in tokamak disruptions. *Phys. Plasmas* **2006**, *13*, 102502. [CrossRef]

79. Minashin, P.V.; Kukushkin, A.B.; Poznyak, V.I. Reconstruction of superthermal electron velocity distribution function from electron cyclotron spectra at down-shifted frequencies in tokamak T-10. *Eur. Phys. J. Conf.* **2012**, *32*, 01015. [CrossRef]

80. Kurzan, B.; Steuer, K.-H.; Suttrop, W. Runaway electrons in a tokamak: A free-electron maser. *Rev. Sci. Instrum.* **1997**, *68*, 423–426. [CrossRef]

81. Ide, S.; Ogura, K.; Tanaka, H.; Iida, M.; Hanada, K.; Itoh, T.; Iwamasa, M.; Sakakibara, H.; Minami, T.; Yoshida, M.; et al. Investigation of high energy electrons in lower hybrid current drive plasma with electron cyclotron emission measurement in the WT-3 tokamak. *Nucl. Fusion* **1989**, *29*, 1325–1338. [CrossRef]

82. Oks, E.; Sanders, P. Stark broadening of hydrogen/deuterium spectral lines by a relativistic electron beam: Analytical results and possible applications to magnetic fusion edge plasmas. *J. Phys. Commun.* **2018**, *2*, 015030. [CrossRef]

83. Rosato, J.; Pandya, S.P.; Logeais, C.; Meireni, M.; Hannachi, I.; Reichle, R.; Barnsley, R.; Marandet, Y.; Stamm, R. A study for Stark broadening for the diagnostic of runaway electrons in ITER. *AIP Conf. Proc.* **2017**, *1811*, 110001. [CrossRef]

84. Ispolatov, Y.; Oks, E. A convergent theory of Stark broadening of hydrogen lines in dense plasmas. *J. Quant. Spectrosc. Radiat. Transf.* **1994**, *51*, 129–138. [CrossRef]

85. Krasovitskiy, V.B. *Instabilities of Relativistic Electron Beam in Plasma*; Nova Publishers: New York, NY, USA, 2008.

86. Gavrilenko, V.P.; Oks, E. A new effect in the Stark spectroscopy of atomic Hydrogen: Dynamic resonance. *Sov. Phys. JETP* **1981**, *53*, 1122–1127.

87. Dalimier, E.; Oks, E.; Renner, O. Review of Langmuir-wave-caused dips and charge-exchange-caused dips in spectral lines from plasmas and their applications. *Atoms* **2014**, *2*, 178–194. [CrossRef]

88. Dalimier, E.; Faenov, A.Y.; Oks, E.; Angelo, P.; Pikuz, T.A.; Fukuda, Y.; Andreev, A.; Koga, J.; Sakaki, H.; Kotaki, H.; et al. X-ray spectroscopy of super-intense laser-produced plasmas for the study of nonlinear processes. Comparison with PIC simulations. *J. Phys. Conf. Ser.* **2017**, *810*, 012004. [CrossRef]

89. Dalimier, E.; Oks, E.; Renner, O. Dips in spectral line profiles and their applications in plasma physics and atomic physics. *AIP Conf. Proc.* **2017**, *1811*. [CrossRef]

90. Oks, E.; Böddeker, S.; Kunze, H.-J. Spectroscopy of atomic hydrogen in dense plasmas in the presence of dynamic fields: Intra-Stark spectroscopy. *Phys. Rev. A* **1991**, *44*, 8338–8347. [CrossRef] [PubMed]

91. Renner, O.; Dalimier, E.; Oks, E.; Krasniqi, F.; Dufour, E.; Schott, R.; Foerster, E. Experimental evidence of Langmuir-wave-caused features in spectral lines of laser-produced plasmas. *J. Quant. Spectr. Radiat. Transf.* **2006**, *99*, 439–450. [CrossRef]

92. Oks, E.; Dalimier, E.; Faenov, A.Y.; Pikuz, T.; Fukuda, Y.; Jinno, S.; Sakaki, H.; Kotaki, H.; Pirozhkov, A.; Hayashi, Y.; et al. Two-plasmon decay instability's signature in spectral lines and spectroscopic measurements of charge exchange rate in a femtosecond laser-driven cluster-based plasma. Fast Track Communications. *J. Phys. B* **2014**, *47*, 221001. [CrossRef]

93. Oks, E.; Dalimier, E.; Faenov, A.Y.; Angelo, P.; Pikuz, S.A.; Tubman, E.; Butler, N.M.H.; Dance, R.J.; Pikuz, T.A.; Skobelev, I.Y.; et al. Using X-ray spectroscopy of relativistic laser plasma interaction to reveal parametric decay instabilities: A modeling tool for astrophysics. *Opt. Express* **2017**, *25*, 1958–1972. [CrossRef] [PubMed]

94. Wagner, U.; Tatarakis, M.; Gopal, A.; Beg, F.N.; Clark, E.L.; Dangor, A.E.; Evans, R.G.; Haines, M.G.; Mangles, S.P.D.; Norreys, P.A.; et al. Laboratory measurements of 0.7 GG magnetic fields generated during high intensity laser interactions with dense plasmas. *Phys. Rev. E* **2004**, *70*. [CrossRef] [PubMed]

95. Fujioka, S.; Zhang, Z.; Ishihara, K.; Shigemori, K.; Hironaka, Y.; Johzaki, T.; Sunahara, A.; Yamamoto, N.; Nakashima, H.; Watanabe, T.; et al. Kilotesla magnetic field due to a capacitor-coil target driven byhigh power laser. *Sci. Rep.* **2013**, *3*, 1170. [CrossRef] [PubMed]

96. Silvers, L.J. Magnetic fields in astrophysical objects. *Phil. Trans. R. Soc. A* **2008**, *366*, 4453–4464. [CrossRef] [PubMed]

97. Schekochihin, A.A.; Cowley, S.C.; Yousef, T.A. MHD turbulence: Nonlocal, anisotropic, nonuniversal? In *Magnetohydrodynamics—Historical Evolution and Trends*; Molokov, S., Moreau, R., Moffett, H.K., Eds.; Springer: Berlin, Germany, 2007.

98. Oks, E. Influence of magnetic-field-caused modifications of trajectories of plasma electrons on spectral line shapes: Applications to magnetic fusion and white dwarfs. *J. Quant. Spectrosc. Radiat. Transf.* **2016**, *171*, 15–27. [CrossRef]

99. Tremblay, P.-E.; Bergeron, P. Spectroscopic analysis of DA white dwarfs: Stark broadening of hydrogen lines including non-ideal effects. *Astrophys. J.* **2009**, *696*, 1755–1770. [CrossRef]

100. Oks, E.; Rantsev-Kartinov, V.A. Spectroscopic observation and analysis of plasma turbulence in a Z-pinch. *Sov. Phys. JETP* **1980**, *52*, 50–58.
101. Oks, E. Effect of helical trajectories of electrons in strongly magnetized plasmas on the width of hydrogen/deuterium spectral lines: Analytical results and applications to white dwarfs. *Int. Rev. At. Mol. Phys.* **2017**, *8*, 61–72.
102. Oks, E. Latest advances in the semiclassical theory of the Stark broadening of spectral lines in plasmas. *J. Phys. Conf. Ser.* **2017**, *810*, 012006. [CrossRef]
103. Oks, E. Refinement of the semiclassical theory of the Stark broadening of hydrogen spectral lines in plasmas. *J. Quant. Spectrosc. Radiat. Transf.* **2015**, *152*, 74–83. [CrossRef]
104. Franzon, B.; Schramm, S. Effect of the magnetic field in white dwards. *J. Phys. Conf. Ser.* **2017**, *861*, 012015. [CrossRef]
105. Reimers, D.; Jordan, S.; Koester, D.; Bade, N.; Köhler, T.; Wisotzki, L. Discoveryof four white dwarfs with strong magnetic fields by the Hamburg/ESO survey. *Astron. Astrophys.* **1996**, *311*, 572–578.
106. Rosato, J.; Ferri, S.; Stamm, R. Influence of helical trajectories of perturbers on Stark line shapes in magnetized plasmas. *Atoms* **2018**, *6*, 12. [CrossRef]
107. Rosato, J.; Capes, H.; Godbert-Mouret, L.; Koubiti, M.; Marandet, Y.; Stamm, R. Accuracy of impact broadening models in low-density magnetized hydrogen plasmas. *J. Phys. B* **2012**, *45*, 165701. [CrossRef]
108. Sholin, G.V.; Demura, A.V.; Lisitsa, V.S. Theory of Stark broadening of hydrogen lines in plasmas. *Sov. Phys. JETP* **1973**, *37*, 1057–1065.
109. Kielkopf, J.F.; Allard, N.F. Shift and width of the Balmer series Hα line at high electron density in a laser-produced plasma. *J. Phys. B* **2014**, *47*, 155701. [CrossRef]
110. Büscher, S.; Wrubel, T.; Ferri, S.; Kunze, H.-J. The Stark width and shift of the hydrogen Hα line. *J. Phys. B* **2002**, *35*, 2889–2897. [CrossRef]
111. Böddeker, S.; Günter, S.; Könies, A.; Hitzschke, L.; Kunze, H.-J. Shift and width of the H_{α} line of hydrogen in dense plasmas. *Phys. Rev. E* **1993**, *47*, 2785–2791. [CrossRef]
112. Flih, S.A.; Oks, E.; Vitel, Y. Comparison of the Stark widths and shifts of the H-alpha line measured in a flash tube plasma with theoretical results. *J. Phys. B* **2003**, *36*, 283–296. [CrossRef]
113. Gigosos, M.A.; Cardenoso, V. New plasma diagnosis tables of hydrogen Stark broadening including ion dynamics. *J. Phys. B* **1996**, *29*, 4795–4838. [CrossRef]
114. Gigosos, M.A. Stark broadening models for plasma diagnostics. *J. Phys. D* **2014**, *47*, 343001. [CrossRef]
115. Oks, E. Effect of thermal collective modes on the Stark broadening of hydrogen spectral lines in strongly coupled plasmas. *J. Phys. B* **2016**, *49*, 065701. [CrossRef]
116. Antonov, A.S.; Zinov'ev, O.A.; Rusanov, V.D.; Titov, A.V. Broadening of hydrogen spectral lines during turbulent heating of a plasma. *Sov. Phys. JETP* **1970**, *31*, 838–839.
117. Zagorodnikov, S.P.; Smolkin, G.E.; Striganova, E.A.; Sholin, G.V. Determination of the turbulence level in a collisionless magnetosonic shock wave by measuring the Stark broadening of the Balmer Hβ line. *JETP Lett.* **1970**, *11*, 323–326.
118. Zagorodnikov, S.P.; Smolkin, G.E.; Striganova, E.A.; Sholin, G.V. A method of measurement of nonequilibrium electric fields in turbulent plasma based on Stark broadening of hydrogen spectral lines. *Sov. Phys. Dokl.* **1970**, *195*, 1065–1068.
119. Zavojskij, E.K.; Kalinin, Y.G.; Skorjupin, V.A.; Shapkin, V.V.; Sholin, G.V. Measurement of electric fields in turbulent plasma based on Stark broadening of hydrogen spectral lines. *Sov. Phys. Dokl.* **1970**, *194*, 55–58.
120. Zavojskij, E.K.; Kalinin, Y.G.; Skorjupin, V.A.; Shapkin, V.V.; Sholin, G.V. Observation of asymmetry of the distribution of turbulent electric fields in a direct discharge plasma by means of the polarization in the Stark profile of the Hα line. *JETP Lett.* **1970**, *13*, 12–14.
121. Levine, M.A.; Gallagher, C.C. Stark broadening for turbulence studies in a confined plasma. *Phys. Lett. A* **1970**, *32*, 14–15. [CrossRef]
122. Ben-Yosef, N.; Rubin, A.G. Optical investigations of electrostatic turbulence in plasmas. *Phys. Lett.* **1970**, *A33*, 222–223. [CrossRef]
123. Zakatov, I.P.; Plakhov, A.G.; Shapkin, V.V.; Sholin, G.V. Measurement of noise level of Langmuir oscillations in the plasma—Beam system according to Stark broadening of hydrogen spectral lines. *Sov. Phys. Dokl.* **1971**, *198*, 1306–1309.

124. Berezin, A.B.; Dubovoj, A.V.; Ljublin, B.V. *Sov. Phys. Tech. Phys.* **1971**, *41*, 2323.

125. Babykin, M.V.; Zhuzhunashvili, A.I.; Oks, E.; Shapkin, V.V.; Sholin, G.V. Polarization spectroscopic analysis of noise produced in a turbulent plasma upon annihilation of oppositely moving magnetic fields. *Sov. Phys. JETP* **1974**, *38*, 86–92.

126. Volkov, Y.F.; Djatlov, V.G.; Mitina, A.I. *Sov. Phys. Tech. Phys.* **1974**, *44*, 1448.

127. Zhuzhunashvili, A.I.; Oks, E. Technique of optical polarization measurements of the plasma Langmuir turbulence spectrum. *Sov. Phys. JETP* **1977**, *46*, 1122–1132.

128. Berezin, A.B.; Ljublin, B.V.; Jakovlev, D.G. *Sov. Phys. Tech. Phys.* **1983**, *53*, 642.

129. Koval, A.N.; Oks, E. Some results of searching for low-frequency plasma turbulence in large chromospheric flares. *Bull. Crimean Astrophys. Obs.* **1983**, *67*, 78–89.

130. Bohm, D.; Pines, D. A collective description of electron interactions: III. Coulomb interactions in a degenerate electron gas. *Phys. Rev.* **1953**, *92*, 609–625. [CrossRef]

131. Oks, E.; Sholin, G.V. On Stark profiles of hydrogen lines in a plasma with low-frequency turbulence. *Sov. Phys. Tech. Phys.* **1977**, *21*, 144–151.

132. Iglesias, C.A.; DeWitt, H.E.; Lebowitz, J.L.; MacGowan, D.; Hubbard, W.B. Low-frequency electric microfield distributions in plasmas. *Phys. Rev. A* **1985**, *31*, 1698–1702. [CrossRef]

133. Oks, E.; Sholin, G.V. On Stark profiles of hydrogen spectral lines in a plasma with Langmuir Turbulence. *Sov. Phys. JETP* **1975**, *41*, 482–490.

134. Huba, J.D. *NRL Plasma Formulary*; Naval Research Laboratory: Washington, DC, USA, 2013.

135. Bellan, P.M. *Fundamentals of Plasma Physics*; Cambridge University Press: Cambridge, UK, 2006.

136. Kadomtsev, B.B. *Collective Phenomena in Plasma*; Nauka: Moscow, Russia, 1988.

137. Spitzer, L. Stark-effect broadening of hydrogen lines. *Phys. Rev.* **1939**, *55*, 699–708. [CrossRef]

138. Spitzer, L. Stark-effect broadening of hydrogen lines. II. Observable profiles. *Phys. Rev.* **1939**, *56*, 39–47. [CrossRef]

139. Spitzer, L. Impact broadening of spectral lines. *Phys. Rev.* **1940**, *58*, 348–357. [CrossRef]

140. Ishimura, T. Stark effect of the Lymann alpha line by a rotating electric field. *J. Phys. Soc. Jpn.* **1967**, *23*, 422–429. [CrossRef]

141. Lisitsa, V.S. Hydrogen atom in a rotating electric field. *Opt. Spectrosc.* **1971**, *31*, 468–469.

142. Fock, V. Zur Theorie des Wasserstoffatoms. *Z. Phys.* **1935**, *98*, 145–154. [CrossRef]

143. Hughes, J.W.B. Stark states and O(4) symmetry of hydrogenic atoms. *Proc. Phys. Soc.* **1967**, *91*, 810–818. [CrossRef]

144. Goldstein, H. More on the prehistory of the Laplace or Runge-Lenz vector. *Am. J. Phys.* **1976**, *44*, 1123–1124. [CrossRef]

145. Demkov, Y.N.; Monozon, B.S.; Ostrovsky, V.N. Energy levels of a hydrogen atom in crossed electric and magnetic fields. *Sov. Phys. JETP* **1970**, *30*, 775–776.

146. Lisita, V.S.; Sholin, G.V. Exact solution of the problem of the broadening of the hydrogen spectral lines in the one-electron theory. *Sov. Phys. JETP* **1972**, *34*, 484–489.

147. Greene, R.L.; Cooper, J.; Smith, E.W. A unified theory of Stark broadening for hydrogenic ions—I: A general theory (including time ordering). *J. Quant. Spectrosc. Radiat. Transf.* **1975**, *15*, 1025–1036. [CrossRef]

148. Derevianko, A.; Oks, E. Exact solution for the impact broadening of hydrogen spectral lines. In *Physics of Strongly Coupled Plasmas*; Kraeft, W.D., Schlanges, M., Eds.; World Scientific: Singapore, 1996.

149. Derevianko, A.; Oks, E. Exact solution for the impact broadening of the hydrogen lines Lyman-beta and Lyman-gamma. *AIP Conf. Proc.* **1999**, *467*, 148–149.

150. Oks, E. The shape of spectral lines of two-electron Rydberg atoms/ions: Analytical solution. *J. Phys. B* **2017**, *50*, 115001. [CrossRef]

151. Nikitin, S.I.; Ostrovsky, V.N. The symmetry of the electron-electron interaction operator in the dipole approximation. *J. Phys. B* **1978**, *11*, 1681–1694. [CrossRef]

152. Landau, L.D.; Lifshitz, E.M. *Mechanics*; Pergamon: Oxford, UK, 1960.

153. Born, M. *Vorlesungen Über Atommechanik*; Springer: Berlin, Germany, 1925.

154. Kotkin, G.L.; Serbo, V.G. *Sbornik Zadach po Klassicheskoj Mechanike (Collection of Problems in Classical Mechanics)*, 2nd ed.; Nauka: Moscow, Russia, 1977.

155. Kotkin, G.L.; Serbo, V.G. *Collection of Problems in Classical Mechanics*, 1st ed.; Pergamon: Oxford, UK, 1971.
156. Sobel'man, I.I. *An Introduction to the Theory of Atomic Spectra*; Pergamon: Oxford, UK, 1972.
157. Wigner, E.P. *Group Theory and Its Application to the Quantum Mechanics of Atomic Spectra*; Academic Press: New York, NY, USA, 1959.

atoms

MDPI

Review

Laboratory Hydrogen-Beta Emission Spectroscopy for Analysis of Astrophysical White Dwarf Spectra

Christian G. Parigger [1],*, Kyle A. Drake [1], Christopher M. Helstern [1] and Ghaneshwar Gautam [2]

[1] Physics and Astronomy Department, University of Tennessee Space Institute, University of Tennessee, Center for Laser Applications, 411 B.H. Goethert Parkway, Tullahoma, TN 37388-9700, USA; kdrake5@me.com (K.A.D.); chris.helstern@gmail.com (C.M.H.)

[2] Physics Department, Fort Peck Community College, 605 Indian Avenue, Poplar, MT 59255, USA; ggautam@fpcc.edu

* Correspondence: cparigge@tennessee.edu; Tel.: +1-931-841-5690

Received: 29 May 2018; Accepted: 28 June 2018; Published: 1 July 2018

Abstract: This work communicates a review on Balmer series hydrogen beta line measurements and applications for analysis of white dwarf stars. Laser-induced plasma investigations explore electron density and temperature ranges comparable to white dwarf star signatures such as Sirius B, the companion to the brightest star observable from the earth. Spectral line shape characteristics of the hydrogen beta line include width, peak separation, and central dip-shift, thereby providing three indicators for electron density measurements. The hydrogen alpha line shows two primary line-profile parameters for electron density determination, namely, width and shift. Both Boltzmann plot and line-to-continuum ratios yield temperature. The line-shifts recorded with temporally- and spatially-resolved optical emission spectroscopy of hydrogen plasma in laboratory settings can be larger than gravitational redshifts that occur in absorption spectra from radiating white dwarfs. Published astrophysical spectra display significantly diminished Stark or pressure broadening contributions to red-shifted atomic lines. Gravitational redshifts allow one to assess the ratio of mass and radius of these stars, and, subsequently, the mass from cooling models.

Keywords: white dwarfs; burning in stars; plasma diagnostics; atomic spectra; plasma spectroscopy; laser spectroscopy; laser-induced breakdown spectroscopy

PACS: 97.20.Rp; 26.20.Cd; 52.70.-m; 32.30-r; 52.25.Jm; 42.62.Fi

1. Introduction

The investigation of spectral characteristics from stellar objects leads to an understanding of their characteristics. Measurement of sun spectra allows one to infer surface temperature. Hydrogen Balmer series absorption spectra are communicated in Rowland tables [1,2] that map the sun. The equivalent widths [3] of the hydrogen alpha line, H_α, hydrogen beta line, H_β, and hydrogen gamma line, H_γ from the sun amount to 0.402 nm, 0.368 nm, and 0.286 nm, respectively. White dwarf spectra that show hydrogen lines are designated as DA stars. The white dwarf (WD) companion to Sirius A, designated as Alpha Canis Majoris B (α CMa B) [4], reveals significant hydrogen spectra at a temperature of 26 kK [5]. For comparison with the sun spectra, H_γ of Sirius B shows an equivalent width [6] of 3.1 nm.

Laboratory measurements with so-called optical emission spectroscopy in a high-current arc [7,8] suggest that observed gravitational redshifts in WD spectra [9,10] may require corrections due to contributions from Stark-effect caused redshifts [11]. Radiative-transfer considerations [12] however confirm that the Stark or pressure shifts have no material or substantial effect on the measurement of gravitational redshifts. With an observed redshift uncertainty between 5 and 10 km/s [12], the pressure shifts are smaller than the error margins. A recent discussion [13] elaborates on the computation of Stark line profiles in successive layers of an WD atmosphere, and concludes that the Stark effect does not cause substantial redshifts.

Laser-induced breakdown spectroscopy [14,15] measures optical emission from the micro-plasma, and the first few Balmer series hydrogen lines can be utilized to determine electron density and temperature [16–25]. The generalized theory of Stark broadening in the analysis of laser-induced plasma experiments [16,17] show hydrogen beta, H_β, profiles at densities of up to $\sim 8 \times 10^{17} cm^{-3}$. For higher densities and from a theory point-of-view, quadrupole and higher order effects need to be considered [18–20]. Extensive hydrogen modeling [21], including recent systematic experimental efforts [22,23], and detailed analysis of Balmer series lines including the H_β peak separation [24,25] reflect the level of knowledge about laboratory plasma. In view of astrophysical WD absorption data [5], laboratory spectra and analysis are essential for determination of astrophysical WD characterization. Experiments utilize frequently other macroscopic measurement methods including photography [26]. H_β is of interest due to providing better accuracy [27–29] than H_α for an electron density of the order of 10^{17} cm^{-3}.

The analysis of measurements and the theory of Balmer-series hydrogen lines and broadening phenomena [27–37] show significant progress towards understanding of a variety of laboratory and astrophysical conditions. Plasma spectroscopy research [27–31] and extensive line shape considerations [32–37] show substantial efforts towards collecting experimental evidence and computational modeling of hydrogen lines. Plasma emission spectra recorded in the laboratory [8] are analyzed, and subsequently, absorption spectra are determined to fit astrophysical WD spectra [38–40]. Laser-induced breakdown in a laboratory cell [17,22,23] can generate conditions suitable for developing diagnosis of DA type white dwarfs.

A generally accepted method for the determination of the mass of WDs is based on gravitational redshifts [10]. In this approach [41], the mean gravitational redshift, v_g, is determined first to find the mass-radius ratio,

$$v_g = c \, \frac{\Delta\lambda}{\lambda} = \frac{G}{c}\left(\frac{M}{R}\right),$$ (1)

where c and G are the speed of light and gravitational constant, respectively. The symbols M and R indicate, respectively, the mass and radius of the WD. The redshift, $\Delta\lambda$, at the wavelength, λ, of the selected line is usually extracted by fitting a line shape to recorded absorption profiles.

The gravitational redshift is commonly expressed in units of km/s, with WD mass, $M_\odot$, and radius, $R_\odot$, in solar units,

$$v_g[km/s] = 0.636 \left(\frac{M_\odot}{R_\odot}\right).$$ (2)

From effective temperature and from evolutionary cooling models [42], the average mass of 449 non-binary DA stars [41] equals 0.65 times the mass of the sun. Details of WD spectra are further investigated by utilizing sufficiently bright light sources for direct measurement of plasma absorption spectra [43,44].

Regarding Sirius B, analysis of Extreme Ultraviolet Explorer data [45] concludes with a set of parameters to describe this white dwarf Sirius A companion. Table 1 displays the inferred [45] important characteristics, namely, temperature, gravity, mass, and radius.

Table 1. Sirius B parameters [45].

Temperature [K]	Gravity [cm/s^2]	Mass$_{Sirius\ B}$/Mass$_{sun}$	Radius$_{Sirius\ B}$/Radius$_{sun}$
$24{,}790 \pm 100$	$\log(g) = 8.57 \pm 0.06$	0.984 ± 0.074	0.0084 ± 0.00025

The estimate of the electron density, $N_{H\,I}$, of the interstellar hydrogen, H I column to Sirius B, is usually quoted [45] as a logarithm, $\log{(N_{H\,I})} = 17.72 \pm 0.1$, or $N_{H\,I} = (5.25 \pm 1.25) \times 10^{17} \text{cm}^{-2}$.

The Sirius B gravitational redshift, $v_{\text{Sirius B}}$, value is according to previous measurement of primarily H_α [9], $v_{\text{Sirius B}} = 89 \pm 16$ km/s. A gravitational redshift of 89 km/s implies a wavelength-shift of 0.144 nm at H_β, and 0.195 nm at H_α.

Measurements from the *Hipparcos* satellite, operated between 1989 and 1993 by the European Space Agency, indicate slightly different parameters for Sirius B when compared to the data in Table 1. Table 2 summarizes the measurements for Sirius B and Procyon B based on *Hipparcos* parallaxes [46].

Table 2. Sirius B and Procyon B masses and radii [46], M/R ratio, and computed gravitational redshifts using Equation (2).

White Dwarf Star	Mass/Mass$_{\text{sun}}$	Radius/Radius$_{\text{sun}}$	M/R	v_g[km/s]
Sirius B	1.03 ± 0.015	0.0111 ± 0.0007	92.79	59.02
Procyon B	0.594 ± 0.012	0.0096 ± 0.0005	61.88	39.35

Knowledge of the WD mass of course is important to determine whether the mass is close to the Chandrasekhar limit of 1.44× solar mass [47]. Additionally, a WD in a binary star configuration that exists within the Roche limit [48] of the larger companion may contribute sufficient mass to lead to a supernova.

Typical sizes of WDs are comparable to the earth but with a mass similar to that of the sun. Investigations of H_β widths, peak separations, and central dip-shifts in the laboratory further elucidate an understanding of white dwarf stars. This work focuses on various aspects of hydrogen Balmer series measurements of laser-induced plasma, including recently communicated findings in the laboratory [22,23] for conditions encountered for DA type WDs. Laser-induced plasma experiments consistently confirm H_β central dip-shifts.

2. Results

The laboratory results of the hydrogen beta line, H_β, and hydrogen alpha line, H_α, of the Balmer series are reviewed first. Selected data from the WD Montreal data base [5] are communicated along with approaches that are common in the astrophysics communities. However, considering the vast variety of scientific papers devoted to the study of astrophysical objects in the optical region, this review presents aspects of laboratory H_β spectroscopy and its application to white dwarf stars.

2.1. Laboratory Experiments

Line-of-sight and radially resolved data, obtained by Abel inversion, are modeled using computed asymmetric profiles [34]. Detailed experiments on H_β emission spectroscopy [23] explore the radial distribution of laser-induced plasma in hydrogen gas.

The determination of electron density, N_e, from the hydrogen beta line frequently employs well-established empirical identities [21,24,25],

$$N_e[\text{cm}^{-3}] = \left[\frac{\Delta w_{H_\beta}[\text{nm}]}{4.8} \right]^{1.46808} \times 10^{17}, \tag{3}$$

$$\log{\left(N_e[\text{cm}^{-3}] \right)} = 16.661 + 1.416 \log{\left(\Delta\lambda_{ps}[\text{nm}] \right)}, \tag{4}$$

or utilizes results from generalized line shape theory [16] in analysis of laser-induced laboratory plasma [17]. The fitted asymmetric H_β profiles indicate central dip-shifts. A systematic study [22] confirms H_β central wavelength shifts and relates the central dip-shift [22,49,50], $\Delta\delta_{ds}$, to electron density, N_e,

$$\Delta\delta_{ds}[\text{nm}] = 0.14 \left(\frac{N_e[\text{cm}^{-3}]}{10^{17}} \right)^{0.67 \pm 0.03}. \tag{5}$$

The empirical formulae for determination of electron density are summarized in Appendix A. Typical H_β and H_α spectra in Appendix B illustrate the central dip-shift and widths. Recently communicated central dip-shifts [22] apply for the N_e range of $(2–20) \times 10^{17}\text{cm}^{-3}$, thereby extending the previous electron density range for Equation (5), obtained from arc and electromagnetically-driven shock tube results [49,50] that are communicated in graphical representation of central dip-shift versus electron density.

For different time delays, τ, from plasma initiation, Figures 1 and 2a display recorded and fitted line-of-sight and Abel-inverted spectra for hydrogen gas in a cell at a pressure of 1.06×10^5 Pa, and Figure 2b shows the center H_β portion at a cell pressure of 1.32×10^5 Pa.

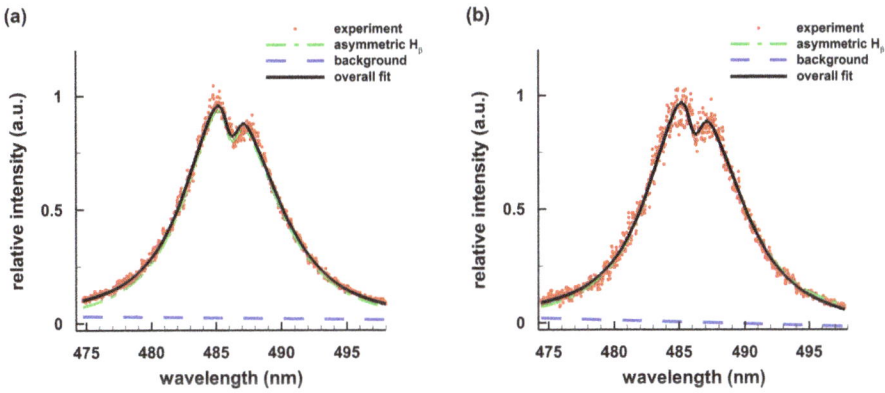

Figure 1. H_β spectra for $\tau = 400$ ns and inferred electron densities of 2.3×10^{17} cm^{-3}: (**a**) line-of-sight at the vertical center of the 4-mm plasma; and (**b**) Abel-inverted spectrum 0.5 mm from the horizontal center [23].

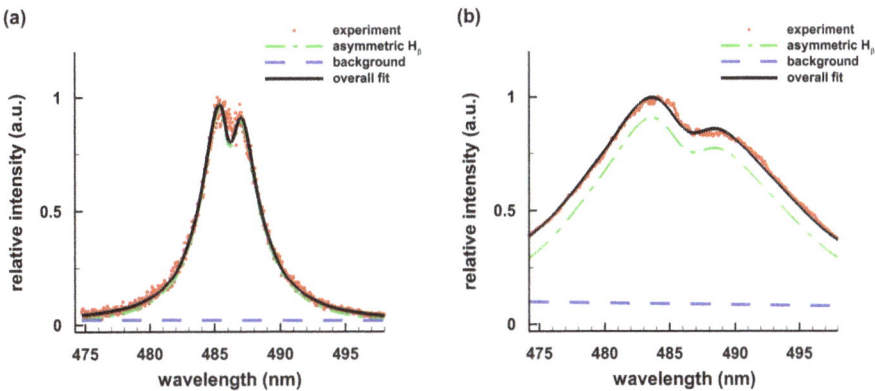

Figure 2. H_β line-of-sight spectra: (**a**) $\tau = 650$ ns, $N_e = 1.4 \times 10^{17}$ cm^{-3}, H_2 gas pressure of 1.06×10^5 Pa [23]; and (**b**) $\tau = 100$ ns, spectral resolving power 24,000 or resolution of 0.02 nm, $N_e = 8.6 \times 10^{17}$ cm^{-3}, H_2 gas pressure of 1.32×10^5 Pa [17].

The line shape of the hydrogen beta line, H_β, due to the Stark effect [51] is described by a Holtsmark profile when considering ion broadening only [52]. However, when accounting for various line broadening phenomena [53] such as effects from Debye shielding, ion and electron broadening, and ion–ion correlations [27,28], the hydrogen beta line profile shows modified central dip and Lorentz line shape asymptotic behavior in the wings. However, deviations from Lorentz line shapes may become noticeable due to incomplete collisions that are not addressed in impact broadening models. These incomplete collisions occur precisely for frequency detuning of the order of several line-widths. Moreover, from a theory point of view, there is need to consider quadrupole and higher order effects for electron densities of the order of 10^{18} cm^{-3} [18–20]. Consideration of asymmetries and dip-shifts further alter the H_β line profile. Appendix C illustrates comparisons of experimental laboratory hydrogen alpha data with Holtsmark, Doppler, and Lorentz line shapes.

2.2. Astrophysical White Dwarf Spectra

Figure 3 illustrates spectra from Sirius B and Procyon B, indicating hydrogen Balmer series lines and C_2 Swan molecular spectra in absorption, respectively. The resolving power, R, for the Sirius B spectrum in Figure 3a equals R $\sim$ 555, or a spectral resolution of 0.88 nm that is too coarse for determination of the expected 0.144-nm redshift. Procyon B (α CMi B) is further evolved [54] in its life time, and it shows a temperature of 8 kK. Procyon B is classified as DQZ type WD and reveals molecular spectra of C_2 due to a carbon-rich and metal-rich atmosphere. Sirius B (α CMa B) reveals a temperature of 26 kK [55] and a typical spectrum for DA white dwarfs. These two stars form the so-called Winter Triangle with Betelgeuse of the constellation Orion.

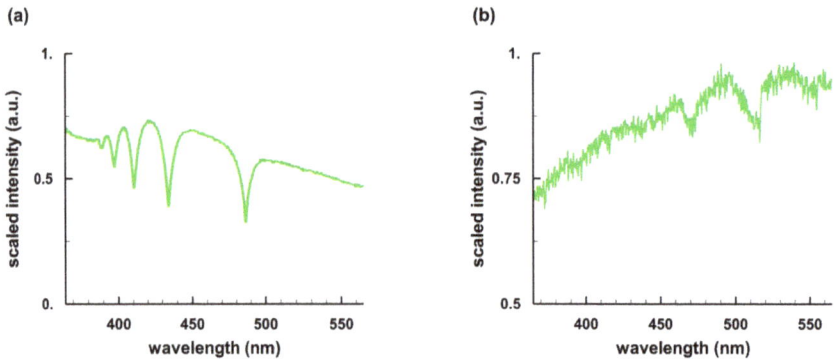

Figure 3. Sirius B and Procyon white dwarf spectra [5]: (**a**) Sirius B at 26 kK; and (**b**) Procyon B at 8 kK.

Figures 4 and 5 display selected DA and magnetic DA (DAH) hydrogen absorption lines [55–57]. Spectra recorded with resolving powers of 800 to 1800 (spectral resolution of 0.61 to 0.27 nm at the H_β 486.14-nm wavelength) rarely if at all show evidence of the hydrogen beta central peak separation. At N_e of 1×10^{17} cm^{-3}, the peak separation and central dip-shift in the laboratory micro-plasma would, respectively, amount to 1.3 nm and 0.14 nm, but again, radiative transfer considerations (e.g., see Reference [13]) indicate that the astrophysical WD spectra do not show significant Stark shifts.

Selected data records of the Sloan digital sky survey [58,59] presents 9316 spectroscopically confirmed white dwarfs, and several WD stars are further analyzed [60]. As an example of a magnetic DA type star (DAH), the Zeeman triplets are nicely recognizable in Figure 5. Magnetic white dwarfs pose challenges [61,62] in the modeling of the recorded absorption spectra. Figure 5 also compares H_β and H_α Zeeman-split and asymmetric Stark-broadened line shapes.

(a)

(b)

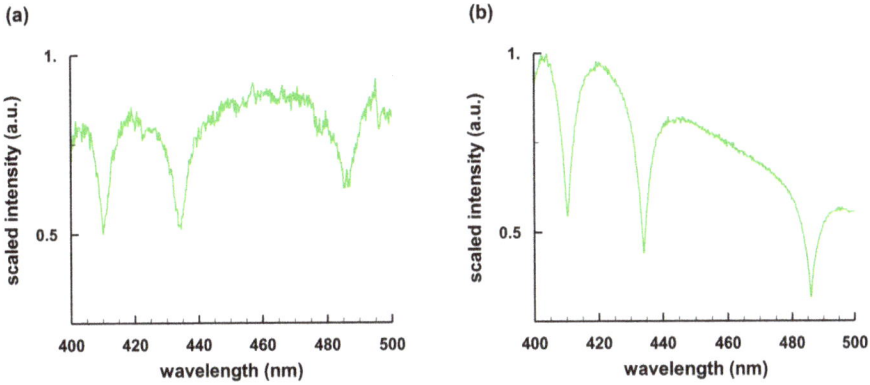

Figure 4. Astrophysical white dwarf spectra: (**a**) WD 1643+143 [55]; and (**b**) WD 1204+023 [5] or SDSS J120650 [57].

(a)

(b)

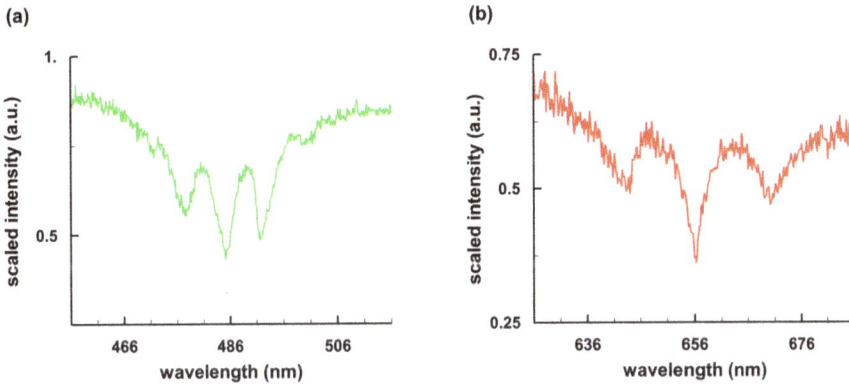

Figure 5. Magnetic white dwarf HS 1031 + 0343 spectra [56]: (**a**) H_β; and (**b**) H_α.

The H_β and H_α profiles in Figure 5 indicate Zeeman-split blue- and red-peak separations, respectively, of $\sigma_\beta = 14.6$ nm and $\sigma_\alpha = 26.6$ nm. The ratio of the H_α peak separation, σ_α, and H_β peak separation, σ_β,

$$\frac{\sigma_\alpha}{\sigma_\beta} = 1.82 = \left(\frac{\lambda_\alpha}{\lambda_\beta}\right)^2, \tag{6}$$

is equal to the square of the ratio of H_α and H_β wavelengths, as expected, although the line shapes appear asymmetric due to Stark broadening. For H_α and H_β, the Zeeman-splits amount to an energy shift 0.038 eV on each side of line center. The shifts in Figure 5 imply magnetic fields of the order of 500 Tesla, also indicated in the computed H_α Zeeman triplet [61] for a magnetic field of 500 Tesla. The electron density estimate equals 3.1×10^{17} cm^{-3}, determined from FWHM of the central absorption of H_β, $\Delta w_{H_\beta} = 10 \pm 1$ nm, and H_α, $\Delta w_{H_\alpha} = 2.7 \pm 0.5$ nm.

Several of the previously discussed white dwarf spectra show a spectral resolution of the order of 1000. However, it would be not unusual to achieve resolving powers of 40,000 (or 0.012 nm at the H_β wavelength) with Echelle spectrometers, for instance, when using the so-called HIRES Echelle spectrometer [63]. The analysis utilizes readily available software, or already extracted sections of an Echelle spectrum. The overall spectral record is composed of parts from different spectral order. Figure 6 shows data for HG 7-85 from the Hyades cluster.

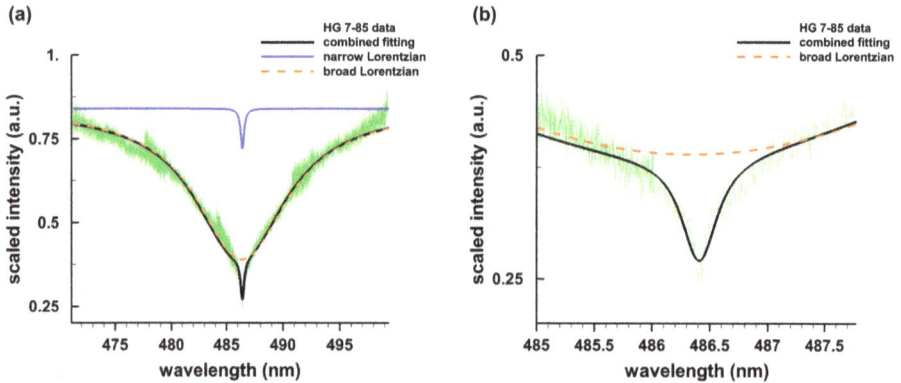

Figure 6. Hyades cluster white dwarf HG 7-85, recorded with a resolving-power 40,000 Echelle-spectrometer: (a) H_β expanded region; and (b) H_β center portion.

The data displayed in Figure 6 are available at KOA [64] following observations [65] on 28–29 October 2012. Figure 6a displays the overlays from Echelle orders, and it shows a broad Lorentzian (dashed) with FWHM of 10 nm, centered at 486.22 nm. Fitting of the composite Echelle spectra is accomplished using publicly available software [66]. The width implies an electron density of 3.1×10^{17} cm^{-3}. The expanded core of the spectrum indicates a shift of 0.3 nm—coincidentally, if time-resolved emission spectroscopy were applied to characterize laboratory plasma, a shift of 0.3 nm (strictly speaking a dip-shift of resolved H_β blue and red peaks) would indicate N_e of 3.1×10^{17} cm^{-3} as well. Figure 6b illustrates a narrow Lorentzian of width 0.19 nm. A FWHM of 0.19 nm would yield an electron density of 0.012×10^{17}cm^{-3}. Comparisons with H_β laboratory results however would suggest that the spectrum is composed of broad and narrow H_β components. The narrow component would imply an electron density that is over two orders of magnitude smaller than that for the broad component.

The gravitational shift of HG 7-85 [65] is 44.3 km/s. Using Equation (1), the corresponding wavelength shift at 486.14 nm amounts to $\Delta\lambda = 0.072$ nm. Note that the center wavelength of the broad Lorentzian fit is at 486.22 nm, or shifted by 0.08 nm . Clearly, further modeling of the atmosphere condition for this Hyades WD would be needed to evaluate the accuracy of this inference. However, the appearance of the H_β spectrum from HG 7-85 could very well indicate absorption from a dense and two orders of magnitude less dense WD atmosphere, as perhaps suggested for other WDs by including carbon, nitrogen, and oxygen in WD atmospheres models [39].

3. Discussion

Broadly speaking, laboratory measurements of the hydrogen Balmer series in the optical region of the electromagnetic spectrum show significant applications in the study of white dwarfs. The research efforts on modeling line shapes extend over several decades if not centuries. At an electron density of 10^{17} cm^{-3}, H_β and H_α widths are well over one order of magnitude larger than Stark-effect redshifts that are measured in laboratory plasma.

However, astrophysical data obtained with increased resolution available at observatories motivate and require accurate measurements of the line shapes and the gravitational redshifts. Accurate laboratory data, modeling and advances in theory are expected to contribute to precise inferences about white dwarf stars, especially for WDs with radiation temperatures in excess of the order of 26 kK or that of Sirius B.

4. Materials and Methods

The experimental arrangement for the laboratory studies include a pulsed, Q-switched, Nd:YAG laser device operated at pulse-widths of 6 to 13 ns using different models. Figure 7 illustrates the schematic for laser-induced plasma experiments inside a laboratory cell.

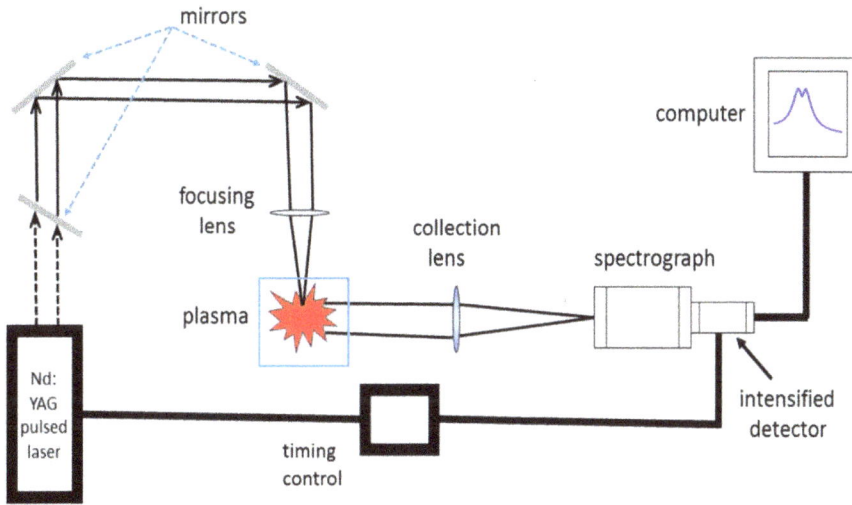

Figure 7. Typical experimental arrangement for generation and measurement of laser-induced plasma.

Laser-induced optical breakdown is accomplished by focusing 150 mJ per pulse of fundamental 1064-nm radiation to achieve of the order of 1000 GW/cm^2 in a cell containing hydrogen gas. Typically, crossed Czerny–Turner spectrometers of 0.25-nm or 0.64-nm focal length disperse the hydrogen lines. In addition, studies of plasma in air or following laser ablation contribute to the set of investigations. A photomultiplier, linear intensified diode array or an intensified charge-coupled device record spectrometer-dispersed light.

Usual sensitivity correction, wavelength calibration, and detector background subtraction deliver data records suitable for analysis that includes de-convolution from the detection system transfer function. An overview video [67] of the measurement protocol provides sufficient detail for laser-induced breakdown spectroscopy for atomic and diatomic molecular analysis [68].

Spectral resolutions for linear diode arrays or ICCDs of the order of 0.1 nm with the 0.64-m Jobin Yvon (JY) and 0.25 nm with the 0.25-m Jarrel–Ash spectrometers, or resolving powers of the order of 5000 and 2000, provide sufficient resolution for the measurement of full-width-half-maximum H$_\beta$ and H$_\alpha$ lines. For the experiments with a photomultiplier and moving a 3600 grooves/mm holographic grating of the JY spectrometer, H$_\beta$ and H$_\alpha$ spectral resolution of 0.02 nm is realized. Measurements of redshifts with diode arrays or ICCDs require sufficient resolving power for electron densities, N$_e$, of nominally 1×10^{17} cm^{-3}. However, redshifts for N$_e$ of 10×10^{17} cm^{-3} are large enough to achieve acceptable error margins.

For H$_\beta$ and H$_\alpha$, the ratio of width to central dip-shift and redshift amounts to 32 and 24, respectively, at 1×10^{17} cm^{-3} with a weak N$_e$ dependency as indicated in Appendix A. Noteworthy, for the experiments with a photomultiplier and moving 3600 grooves/mm grating of the JY spectrometer, the H_α spectral resolution of 0.02 nm is realized, or a resolving power of 33,000 at 656.28 nm. For H_β, spectral resolving powers of the order of 50,000 at 486.14 nm or a resolution of 0.01 nm with the JY spectrometer would in principle be available when using a photomultiplier that preferably would need to be gated for accurate measurements at longer time delays of say 175 ns to block the early and intense plasma radiation as the line emerges from the free electron background radiation.

The analysis of the recorded line profiles, over and above the evaluation of the FWHM, includes fitting of tabulated profiles [27,28], fitting to Voigt profiles or Lorentz profiles, and, recently, fitting to asymmetric hydrogen beta profiles [34]. The latter approach allows one to efficiently analyze Abel-inverted data sets in investigations of radial plasma expansion phenomena.

5. Conclusions

The laboratory laser-induced plasma investigations using temporally- and spatially-resolved spectroscopy clearly indicate direct application to determination of astrophysical white dwarf parameters. Data collection with a resolving power of the order of 50,000 to 75,000 or better appears to be desirable for astrophysical and laboratory spectra to determine accurate hydrogen beta line profile parameters for the encountered white dwarf spectral redshifts and electron densities.

Moreover, the line shape of especially the hydrogen beta line deserves further theoretical attention to exactly reconcile asymmetries and shifts noticeable in laboratory optical emission spectroscopy and in part discernible in recorded astrophysical absorption spectra.

Author Contributions: C.G.P. conceived, designed, and performed the experiments; K.A.D. and G.G. contributed to the analysis of the spectra; K.A.D., G.G., and C.M.H. assisted in extensive literature searches; G.G. assisted in the collection of the data; and C.G.P. outlined and wrote the paper.

Funding: The authors appreciate the support in part by the Center for Laser Application, a State of Tennessee funded Accomplished Center of Excellence at the University of Tennessee Space Institute.

Acknowledgments: This research made use of the Keck Observatory Archive (KOA), which is operated by the W. M. Keck Observatory and the NASA Exoplanet Science Institute (NExScI), under contract with the National Aeronautics and Space Administration. One of us (CGP) thanks P. Dufour for promptly responding with text files of WD spectra listed at the Montreal white dwarf database, SDSS-III Science Archive Scientist Benjamin Alan Weaver for his support with SDSS data, Lasker Data Science Fellow G. Narayan for sending text files of data from the Gemini Observatory and the Multiple Mirror Telescope, and B. Zuckerman and Beth Klein for communication exchanges about their Hyades study.

Conflicts of Interest: The authors declare no conflict of interest. The funding sponsors had no role in the design of the study; in the collection, analyses, or interpretation of data; in the writing of the manuscript, and in the decision to publish the results.

Abbreviations

The following abbreviations are used in this manuscript:

c	speed of light
CMa	Canis Majoris—big dog
CMi	Canis Minoris—little dog
$\Delta w_{H\alpha}$	hydrogen alpha FWHM
$\Delta\lambda_{H\alpha}$	hydrogen alpha redshift
$\Delta w_{H\beta}$	hydrogen beta FWHM
$\Delta\lambda_{ps}$	hydrogen beta peak separation
$\Delta\lambda_{ds}$	hydrogen beta central dip-shift
DA	dwarf A—hydrogen lines are present
DAH	dwarf A with magnetic field
DQZ	dwarf Q and Z—carbon and metal rich atmosphere
FWHM	full width at half maximum
G	gravitational constant

H_β	hydrogen beta line
H_α	hydrogen alpha line
HIRES	high resolution spectrometer
ICCD	intensified charge coupled device
JA	Jarrel-Ash
JY	Jobin Yvon
KOA	Keck observatory archive
λ_α	wavelength of H_α
λ_β	wavelength of H_β
M	WD mass
$M_\odot$	WD mass in solar units
$N_{H\,I}$	electron interstellar density of the hydrogen column
N_e	electron density
R	WD radius
$R_\odot$	WD radius in solar units
σ_α	blue- and red-peak Zeeman separation for H_α
σ_β	blue- and red-peak Zeeman separation for H_β
SDSS	Sloan digital sky survey
τ	time delay from optical breakdown
v_g	gravitational redshift
WD	white dwarf
$\chi(\beta)$	integral describing predominant Holtsmark contribution
$\Lambda(\beta)$	integral describing predominant Lorentz contribution
$\Delta(\beta)$	integral describing predominant Doppler contribution
β	ration of electric and normal field strength
F	electric field strength
F_0	normal field strength
CGP	Christian Gerhard Parigger
CMH	Christopher Matthew Helstern
KAD	Kyle Anthony Drake
GG	Ghaneshwar Gautam

Appendix A. Formulae for Determination of Electron Density from H_β and H_α Profiles

The set of empirical formulae for H_β and H_α derive from various laboratory measurements of laser-induced plasma in gases and laser ablation [69]. Linear log-log fitting yields the formulae for H_β electron densities in the range of 0.03 to 8×10^{17} cm^{-3} for widths of 3 to 20 nm,

$$\Delta w_{H_\beta}[\text{nm}] = 4.5 \left(\frac{N_e[\text{cm}^{-3}]}{10^{17}} \right)^{0.71 \pm 0.03}. \tag{A1}$$

The peak-separation of the two H_β shifted peaks,

$$\Delta\lambda_{ps}[\text{nm}] = 1.3 \left(\frac{N_e[\text{cm}^{-3}]}{10^{17}} \right)^{0.61 \pm 0.03}. \tag{A2}$$

The central dip-shift [22],

$$\Delta\lambda_{ds}[\text{nm}] = 0.14 \left(\frac{N_e[\text{cm}^{-3}]}{10^{17}} \right)^{0.67 \pm 0.03}, \tag{A3}$$

provide further measures of the electron density. For electron densities, N_e, higher than 7×10^{17} cm^{-3}, especially the redshift of the central dip allows one to determine N_e up to 20×10^{17} cm^{-3} [22].

The H_α width and redshift formulae,

$$\Delta w_{H\alpha}[\text{nm}] = 1.3 \left(\frac{N_e[\text{cm}^{-3}]}{10^{17}} \right)^{0.64 \pm 0.03}, \qquad (A4)$$

$$\Delta \lambda_{H\alpha}[\text{nm}] = 0.055 \left(\frac{N_e[\text{cm}^{-3}]}{10^{17}} \right)^{0.97 \pm 0.03}, \qquad (A5)$$

apply to electron densities of the order of 0.01 to 100×10^{17} cm^{-3}. The formulae for electron density determination from H_α are consistent within error margins of reported gas-liner pinch data and diagnosis using Thomson scattering and simultaneous spectroscopic measurements [70].

Electron densities, N_e, determined from H_β widths are preferred due the 5% to 10% accuracy that can be achieved [30]. This is in part due to the larger width for H_β than for H_α indicated in Equations (A1) and (A3), or due to the H_β peak separation that can be utilized for electron density diagnostics [25]. Table A1 displays measured shifts of the central H_β dip and the computed electron density using Equation (A3). The H_β widths are too wide for time delays smaller than 175 ns and for the selected measurement window, but the results are consistent with the H_α widths.

Table A1. Average H_β central dip-shifts, widths, and electron densities, N_e, from dip-shifts and widths for selected time delays of optical breakdown in 0.76×10^5 Pa hydrogen gas [22].

Time Delay (ns)	H_β Dip-Shift (nm)	H_β Width (nm)	N_e from Dip-Shift (10^{17} cm^{-3})	N_e from Width (10^{17} cm^{-3})
25	1.02 ± 0.15	-	20 (15–24)	- (H_α: 17)
50	0.83 ± 0.15	-	14 (11–19)	- (H_α: 14)
75	0.77 ± 0.1	~25	13 (10–15)	~11
100	0.65 ± 0.1	~23 $\pm$ 4	10 (7.7–12)	~10
125	0.58 ± 0.1	~21 $\pm$ 4	8.4 (6.3–11)	~8.8
150	0.50 ± 0.1	~15 $\pm$ 3	6.7 (4.8–8.9)	~5.5
175	0.42 ± 0.1	11 $\pm$ 2	5.2 (3.5–7.2)	3.5
200	0.37 ± 0.1	9 $\pm$ 1	4.3 (2.7–6.1)	2.7
225	0.32 ± 0.1	8 $\pm$ 0.5	3.5 (2.0–5.2)	2.3
250	0.26 ± 0.05	7.5 $\pm$ 0.5	2.5 (1.8–3.3)	2.1
275	0.24 ± 0.05	7.0 $\pm$ 0.5	2.2 (1.6–3.0)	1.9

Previous laser-induced plasma work [17] addresses determination of electron density from full widths at half-maximum of the Balmer series hydrogen-beta line, H_β, peak-separation of H_β, and comparisons with hydrogen-alpha line, H_α, results. Re-analysis in view of the central dip-shifts in Figure 1a,b of Reference [17] shows values of 0.5 ± 0.025 nm and 0.33 ± 0.025 nm, respectively, that lead to N_e of $6.7 \pm 0.5 \times 10^{17}$ cm^{-3} and $3.6 \pm 0.4 \times 10^{17}$ cm^{-3}. These values are consistent with the determined N_e from in the last column of Table 2 in Reference [17].

Appendix B. Typical Laboratory Spectra of H_β and H_α

Recent laboratory experiments [22] reveal central H_β dip-shifts from plasma emissions following optical breakdown in 0.76×10^5 Pa hydrogen gas. Typical spectra of H_β also show the separation of the two red and blue peaks, including the asymmetry. H_α shows central Stark components, therefore, the central dip is absent.

Figures A1 and A2 show H_β and H_α pseudo-colored images for a time delay, τ, of 250 ns from laser-plasma initiation. The figures also show the corresponding averages along the slit. From the FWHM and using formulae in Appendix A, the average electron density is $N_e = 2.1 \times 10^{17}$ cm^{-3} (see Table A1 in Appendix A). The electron temperature, T_e, equals 50 kK (4.3 eV) for $\tau = 250$ ns [22], evaluated from line-to-continuum and Boltzmann plots with an error margin of ± 10 kK (0.86 eV).

(a)   **(b)**

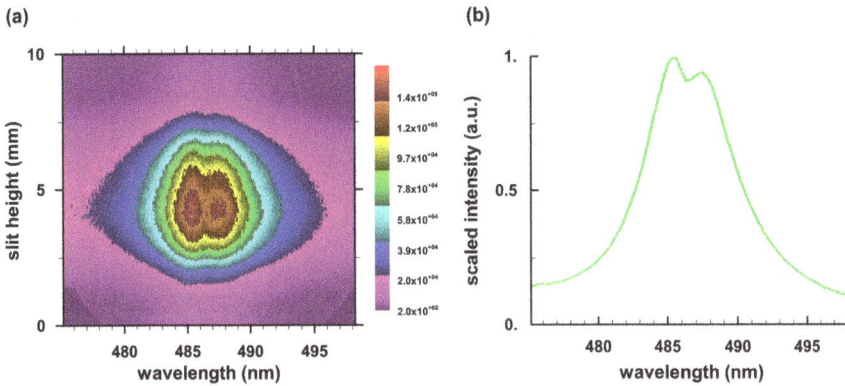

Figure A1. H_β spectra for $\tau = 250$ ns: (**a**) H_β map; and (**b**) scaled average.

(a)   **(b)**

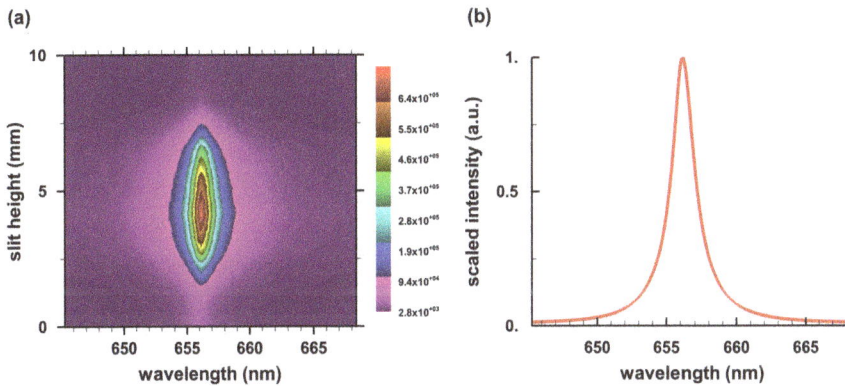

Figure A2. H_α spectra for $\tau = 250$ ns: (**a**) H_α map; and (**b**) scaled average.

In the experiments, the 1064-nm laser beam propagates from the top to the bottom and parallel to the spectrometer slit. The time-resolved records are collected with an ICCD, accumulating 100 consecutive laser-plasma events. The profiles along the slit indicate the average electron density variation along the plasma. The plasma expansion causes higher N_e near the shock wave or the expanding plasma kernel boundary than for the other regions, leading to line-of-sight profiles that appear as a superposition of lower and higher density line shapes, analogous to reported laser ablation spectral line shapes [71,72]. The average along the slit dimension yields spectra similar to data acquired with linear diode arrays.

Appendix C. Line Shapes

The literature indicates extensive material on the computation of hydrogen line profiles. Historically speaking, Schrödinger published his "modern" quantum mechanics calculation of Stark shifts for Balmer series lines H_α, H_β, H_γ, and H_δ [51]. Figure A3 illustrates measured H_β and H_α spectra including overlays of Holtsmark, Lorentz and Doppler profiles of the same widths.

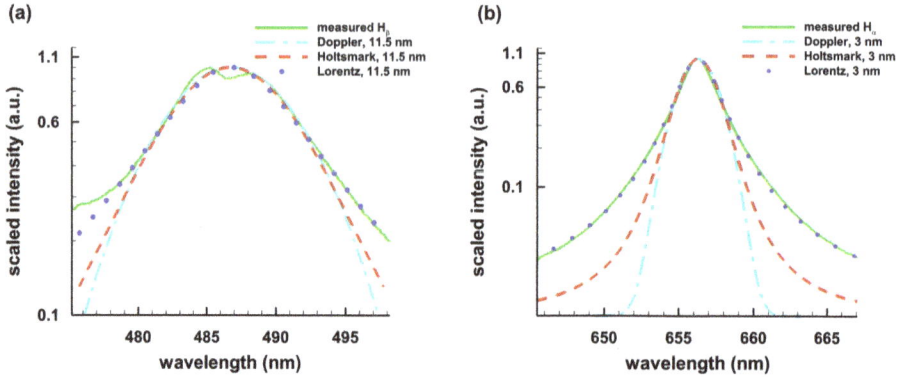

Figure A3. Comparison of measured data at a time delay of 175 ns, one-dimensional Holtsmark, Lorentz, and Doppler line shapes with the same full-width-half-maxima: (a) H_β; and (b) H_α.

Broadening considerations of these lines include results obtained by J. Holtsmark [52] especially in the quasi-static limit for the ions. Doppler or inhomogeneous and Lorentz or homogeneous broadening describe velocity and electron-collision impact phenomena, respectively.

From a mathematical point of view, these three spectral line shapes represent the probability distribution function of the one-dimensional Holtsmark, Lorentz, and Doppler characteristic functions. Table A2 displays the characteristic functions of the independent variable t including mean, μ_n, and scale parameters, c_n, with n = h, l, d.

Table A2. Characteristic functions, $\varphi(x)$ for one-dimensional Holtsmark, Lorentz, and Doppler profiles.

Holtsmark	Lorentz	Doppler
$\exp\{it\mu_h - |c_h t|^{3/2}\}$	$\exp\{it\mu_l - c_l t\}$	$\exp\{it\mu_d - (c_d t)^2\}$

A spectral line profile can be viewed as the standard integral representation of the characteristic function of a continuous distribution defined by its probability distribution function. Computation of the line profiles is based on the Matlab® Characteristic Functions Toolbox [73]. In other words, the spectral line is evaluated as the real part of the Fourier transform as function of x,

$$I(x) = \frac{1}{2\pi} \int_{-\infty}^{+\infty} \varphi(t) \exp\{-ixt\} dt. \tag{A6}$$

For zero mean, $\mu_h = \mu_l = \mu_d = 0$, constant scaling, $c_h = c_l = c_d = 5.75$ and $c_h = c_l = c_d = 1.5$ for H_β and H_α, respectively, and with adjustments for a 1% background offset for Holtsmark, Lorentz, and Doppler profiles with same full-width-half-maxima (FWHM) of 11.5 and 3 nm, Figure A3 shows close to Lorentz profiles in the wings indicating the impact approximation for electrons.

The H_β profiles would indicate Holtsmark profiles in the wings if only ion broadening were considered [27,28], but detailed microfield [74] considerations and Stark broadening computations [27,28,75,76] include electron broadening, Debye shielding, and ion-ion correlations that decrease the depth fidelity at line center, and cause Lorentz-like dependencies in the wings. In laboratory measurements of H_β, the Doppler width is usually smaller (~0.08 nm at 50 kK) than the spectral resolution (~0.1 nm) and much smaller than the FWHM (~11 nm) for a density 3.6×10^{17} cm^{-3}. Computations of line shapes usually involve the convolution of at least the Holtsmark, Lorentz, and Doppler profiles. The convolution of the Holtsmark and Voigt profiles is reduced to a sum of three integrals with predominant Holtsmark, Lorentz, and Doppler contributions [77,78], respectively,

$$\Phi(\beta) = \chi(\beta) + \Lambda(\beta) + \Delta(\beta). \tag{A7}$$

Here, β indicates the ratio of electric field strength, F, and of normal field strength, F_0, $\beta = F/F_0$. The three integrals $\chi(\beta)$, $\Lambda(\beta)$, and $\Delta(\beta)$ are explicitly listed in Reference [78].

The Holtsmark distribution is frequently encountered in the description of astrophysical plasma. In laboratory plasma, the Holtsmark distribution describes ion-ion effects in Stark broadened lines. For the encountered densities of the order of 1 to 10×10^{17} cm^{-3} in the H$_\beta$ measurements, the wings are primarily collision-broadened, described by a Lorentz profile (see Figure A3a). Equally, H$_\alpha$ in Figure A3b reveals Lorentz type wings that indicate electron impact broadening.

References

1. Rowland, H.A. Preliminary Table of the Solar Spectruim Wave-Lengths VI. *Astrophys. J.* **1895**, *2*, 45–54. [CrossRef]
2. Rowland, H.A. Preliminary Table of the Solar Spectruim Wave-Lengths XIII. *Astrophys. J.* **1895**, *3*, 356–360. [CrossRef]
3. Moore, C.E.; Minnaert, M.G.J.; Houtgast, J. *The Solar Spectrum 2935 Å to 8770 Å*; National Burea of Standards Monograph 61; United Styates Department of Commerce: Washington, DC, USA, 1966; pp. 197–283.
4. Barstow, M.A.; Bond, H.E.; Holberg, J.B.; Burleigh, M.R.; Hubeny, I.; Koester, D. Hubble Space Telescope Spectroscopy of the Balmer lines in Sirius B. *Mon. Not. R. Astron. Soc.* **2005**, *362*, 1134–1142. [CrossRef]
5. Dufour, P.; Blouin, S.; Coutu, S.; Fortin-Archambault, M.; Thibeault, C.; Bergeron, P.; Fontaine, G. The Montreal White Dwarf Database: A Tool for the Community. In *Astronomical Society of the Pacific (ASP) Conference Series 509, Proceedings of the 20th European White Dwarf workshop, Warwick, UK, 25–29 July 2016*; Tremblay, P.-E., Gaensicke, B., Marsh, T., Eds.; Utah Valley University: Orem, UT, USA, 2017; pp. 3–8, ISBN 978-1-58381-903-6. Available online: http://dev.montrealwhitedwarfdatabase.org (accessed on 11 May 2018).
6. Kodaira, K. A Spectrum of Sirius B. *Astron. Soc. J.* **1967**, *19*, 172–179.
7. Wiese, W.L.; Kelleher, D.E. On the Cause of the Redshifts in White-Dwarf Spectra. *Astrophys. J.* **1971**, *166*, L59–L63. [CrossRef]
8. Wiese, W.L.; Kelleher, D.E.; Paquette, D.R. Detailed Study of Stark Broadening of Balmer Lines in High-Density Plasma. *Phys. Rev. A* **1972**, *6*, 1132–1153. [CrossRef]
9. Greenstein, J.L.; Oke, J.B.; Shipman, H.L. Effective temperature, radius, and gravitational redshift of Sirius B. *Astrophys. J.* **1971**, *169*, 563–566. [CrossRef]
10. Holberg, J.B. Sirius B and the Measurement of the Gravitational Redshift. *J. His. Astron.* **2010**, *41*, 41–64. [CrossRef]
11. Grabowski, B.; Halenka, J. On red shifts and asymmetries of hydrogen spectral lines. *Astron. Astrophys.* **1975**, *45*, 159–166.
12. Shipman, H.L.; Mehan, R.G. The unimportance of pressure shifts in the measurement of gravitational redshifts in white dwarfs. *Astrophys. J.* **1976**, *209*, 205–207. [CrossRef]
13. Halenka, J.; Olchawa, W.; Madej, J.; Grabowski, B. Pressure shift and gravitational redshift of Balmer lines in white dwarfs: Rediscussion. *Astrophys. J.* **2015**, *808*, 131–140. [CrossRef]
14. Cremers, D.A.; Radziemski, L.J. *Handbook of Laser-Induced Breakdown Spectroscopy*; John Wiley & Sons: Hoboken, NJ, USA, 2006; pp. 1–97, ISBN 978-0-470-09299-6.
15. Hahn, D.W.; Omenetto, N. Laser-Induced Breakdown Spectroscopy (LIBS), Part II: Review of Instrumental and Methodological Approaches to Material Analysis and Applications to Different Fields. *Appl. Spectrosc.* **2012**, *66*, 47–419. [CrossRef] [PubMed]
16. Ispolatov, Y.; Oks, E. A convergent theory of Stark broadening of hydrogen lines in dense plasmas. *J. Quant. Spectrosc. Rad. Trans.* **1994**, *51*, 129–138. [CrossRef]
17. Parigger, C.G.; Plemmons, D.H.; Oks, E. Balmer series H$_\beta$ measurements in a laser-induced hydrogen plasma. *Appl. Opt.* **2003**, *42*, 5992–6000. [CrossRef] [PubMed]

18. Stambulchik, E.; Fisher, D.V.; Maron, Y.; Griem, H.R.; Alexiou, S. Correlation effects and their influence on line broadening in plasmas. *High Energy Density Phys.* **2007**, *3*, 272–277. [CrossRef]
19. Gomes, T.A.; Nagayama, T.; Kilcrease, D.P.; Montgomery, M.H.; Winget, D.E. Effect of higher-order multipole moments on the Stark line shape. *Phys. Rev. A* **2016**, *94*, 022501. [CrossRef]
20. Demura, A.V. Beyond the Linear Stark Effect: A Retrorespective. *Atoms* **2018**, *6*, 33. [CrossRef]
21. Gigosos, M.A.; González ,M.Á.; Cardeñoso, V. Computer simulated Balmer-alpha, -beta and -gamma Stark line profiles for non-equilibrium plasmas diagnostics. *Spectrochim. Acta B At. Spectrosc.* **2003**, *58*, 1489–1504. [CrossRef]
22. Parigger, C.G.; Helstern, C.M.; Drake, K.A.; Gautam, G. Balmer-series hydrogen-beta line dip-shifts for electron density measurements. *Int. Rev. At. Mol. Phys.* **2017**, *8*, 53–61.
23. Parigger, C.G.; Gautam, G.; Surmick, D.M. Radial electron density measurements in laser-induced plasma from Abel inverted hydrogen Balmer beta line profiles. *Int. Rev. At. Mol. Phys.* **2015**, *6*, 43–55.
24. Konjević, N.; Ivković, M.; Sakan, N. Hydrogen Balmer lines for low electron number density plasma diagnostics. *Spectrochim. Acta B At. Spectrosc.* **2012**, *76*, 16–26. [CrossRef]
25. Ivković, M.; Konjević, N.; Pavlović, Z. Hydrogen Balmer beta: The separation between line peaks for plasma electron density diagnostics and self-absorption test. *J. Quant. Spectrosc. Radiat. Trans.* **2015**, *154*, 1–8. [CrossRef]
26. Leonard, S.L. Basic Macroscopic Measurements. In *Plasma Diagnostic Techniques*; Huddlestone, R.H., Leonard, S.L., Eds.; Academic Press: New York, NY, USA, 1965; pp. 7–67.
27. Griem, H.R. *Plasma Spectroscopy*; McGraw-Hill Book Company: New York, NY, USA, 1964.
28. Griem, H.R. *Spectral Line Boradening by Plasmas*; Academic Press: Cambridge, MA, USA, 1974.
29. Kunze, H.-J. *Introduction to Plasma Spectroscopy*; Springer: New York, NY, USA, 2009; ISBN 978-3-642-02232-6.
30. Wiese, W.L. Line Broadening. In *Plasma Diagnostic Techniques*; Huddlestone, R.H., Leonard, S.L., Eds.; Academic Press, New York, NY, USA, 1965; pp. 265–317.
31. McWhirter, R.W.P. Spectral Intensities. In *Plasma Diagnostic Techniques*; Huddlestone, R.H., Leonard, S.L., Eds.; Academic Press: New York, NY, USA, 1965; pp. 201–264.
32. Sobel'man, I.I.; Vainshtein, L.A.; Yukov, E.A. *Ecitation of Atoms and Broadening of Spectral Lines*, 2nd ed.; Springer: New York, NY, USA, 1995; ISBN 978-3-540-58686-9.
33. Fujimoto, T. *Plasma Spectroscopy*; Clarendon Press: Oxford, UK, 2004; pp. 213–235, ISBN 9780198530282.
34. Djurović, S.; Ćirišan, M.; Demura, A.V.; Demchenko, G.V.; Nikolić, D.; Gigosos, M.A.; Gonzáles, M.Á. Measurements of H_β central asymmetry and its analysis through standard theory and computer simulations. *Phys. Rev. E* **2009**, *79*, 046402. [CrossRef] [PubMed]
35. Stambulchick, E.; Maron, Y. Plasma line broadening and computer simulations: A mini-review. *High Energy Density Phys.* **2010**, *6*, 9–14. [CrossRef]
36. Oks, E. *Stark Broadening of Hydrogen and Hydrogenlike Spectral Lines in Plasmas: The Physical Insight*; Alpha Science International: Oxford, UK, 2006; pp. 1–118, ISBN 1-84265-252-4.
37. Oks, E. *Diagnostics of Laboratory and Astrophysical Plasmas Using Spectral Lineshapes of One-, Two-, and Three-Electron Systems*; World Scientific: Singapore, 2017; pp. 1–47, ISBN 9789814699075.
38. Tremblay, P.-E.; Bergeron, P. Spectroscopic Analysis of DA White dwarfs: Stark broadening of hydrogen lines including nonideal effects. *Astrophys. J.* **2009**, *696*, 1755–1770. [CrossRef]
39. Gianninas, A.; Bergeron, P.; Dupuis, J.; Ruiz, M.T. Spectroscopic analysis of hot, hydrogen-rich white dwarf: The presence of metals and the Balmer-line problem. *Astrophys. J.* **2010**, *720*, 581–602. [CrossRef]
40. Tremblay, P.-E. Analyse Spectroscopique D'étoiles Naines Blanches Riches en Hydrogène (DA): Vers des Modèles D'atmosphère Améliorés Sans Paramètres Libres. Ph.D. Thesis, Université de Montéal, Montréal, QC, Canada, 2011.
41. Falcon, R.E.; Winget, D.E.; Montgomery, M.H.; Williams, K.A. A gravitational redshift determination of the mean mass of white dwarf DA stars. *Astrophys. J.* **2010**, *712*, 585–595. [CrossRef]
42. Koester, D. White dwarf spectra and atmosphere models. *Mem. Soc. Astronom. Ital.* **2010**, *81*, 921–931, ISSN 1824-016X.
43. Falcon, R.E. Creating and Measuring White Dwarf Photospheres in a Terrestrial Laboratory. Ph.D. Thesis, University of Texas, Austin, TX, USA, 2014.

44. Falcon, R.E.; Rochau ,G.A.; Bailey, J.E.; Gomez, T.A.; Montgomery, M.H.; Winget, D.E.; Nagayama, T. Laboratory measurements of white dwarf photospheric spectral lines: Hβ. *Astrophys. J.* **2015**, *806*, 214–224. [CrossRef]
45. Holberg, J.B.; Barstow, M.A.; Bruhweiler, F.C.; Cruise, A.M.; Penny, A.J. Sirius B: A new, more accurate view. *Astrophys. J.* **1998**, *497*, 935–942. [CrossRef]
46. Provencal, J.L.; Shipman, H.L.; Høg, E.; Thejll, P. Testing the white dwarf mass-radius relation with *Hipparcos*. *Astrophys. J.* **1998**, *494*, 759–767. [CrossRef]
47. Chandrasekhar, S. The maximum mass of ideal white dwarfs. *Astrophys. J.* **1931**, *74*, 81–82. [CrossRef]
48. King, A.R. Accretion and evolution in close binaries. In *Astrophysical and Laboratory Plasmas—A Festschrift for Professor Robert Wilson*; Willis, A.J., Hartquist, T.W., Eds.; Springer, Dordrecht, NL, USA, 1996; pp. 169–186, ISBN 978-90-481-4729-8.
49. Halenka, J.; Vujičić, B.; Djurović, S. Shift of the peaks of the Hβ spectral line. *J. Quant. Spectrosc. Radiat. Trans.* **1989**, *42*, 571–573. [CrossRef]
50. Mijatović, Z.; Pavlov, M.; Djurović, S. Shift of the Hβ line in dense hydrogen plasmas. *Phys. Rev. A* **1991**, *43*, 6095–6097. [CrossRef] [PubMed]
51. Schrödinger, E. Quantisierung als Eigenwertproblem. *Ann. Phys.* **1926**, *385*, 437–490. [CrossRef]
52. Holtsmark, J. Über die Verbreiterung von Spectrallinien. *Ann. Phys.* **1919**, *58*, 577–630. [CrossRef]
53. Van Regenmorter, H. Spectral line broadening. *Ann. Rev. Astron. Astrophys.* **1965**, *3*, 71–92. [CrossRef]
54. Limoǵe, M.-M.; Bergeron, P.; Lépine, S. Physical properties of the current census of northern white dwarfs within 40 pc of the sun. *Astrophys. J.* **2011**, *219*, 19–53. [CrossRef]
55. Gianninas, A.; Bergeron, P.; Ruiz, M.T. A spectroscopic survey and analysis of bright, hydrogen-rich white dwarfs. *Astrophys. J.* **2011**, *743*, 138–164. [CrossRef]
56. Koester D.; Voss, B.; Napiwotzki, R.; Christlieb, N.; Homeier, D.; Lisker, T.; Reimers, D.; Heber, U. High-resolution UVES/VLT spectra of white dwarfs observed for the ESO SN Ia Progenitor Survey III. DA white dwarfs. *Astron. Astrophys.* **2009**, *505*, 441–462. [CrossRef]
57. Narayan, G.; Axelrod, T.; Holberg, J.B.; Matheson, T.; Saha, A.; Olszewski, E.; Claver, J.; Stubbs, C.W.; Bohlin, R.C.; Deustua, S.; Rest, A. Toward a network of faint DA white dwarfs as high-precision spectrophotometric standards. *Astrophys. J.* **2016**, *822*, 67–80. [CrossRef]
58. Kleinman, S.J.; Harris, H.C.; Eisenstein, D.J.; James Liebert, J.; Nitta, A.; Jurek Krzesiński, J.; Munn, J.A.; Dahn, C.C.; Hawley, S.L.; Pier, J.R.; et al. A catalog of spectroscopically confirmed white dwarfs oin the first data realease of the Sloan digital sky survey. *Astrophys. J.* **2004**, *607*, 426–444. [CrossRef]
59. Eisenstein, D.J.; Liebert, J.; Harris, H.C.; Kleinman, S.J.; Nitta, A.; Nicole Silvestri, N.; Anderson, S.A.; Barentine, J.C.; Brewington, H.J.; Brinkmann, J.; et al. A catalog of spectroscopically confirmed white dwarfs from the Sloan digital sky survey data release 4. *Astrophys. J. Suppl. Ser.* **2006**, *167*, 40–58. [CrossRef]
60. Tremblay, P.-E.; Bergeron, P.; Gianninas, A. An improved spectroscopic analysis of DA white dwarfs from the Sloan digital survey data release 4. *Astrophys. J. Suppl. Ser.* **2011**, *730*, 128–150. [CrossRef]
61. Rosato, J.; Kieu, N.; Hannachi, I.; Koubiti, M.; Marandet Y.; Stamm, R.; Dimitrijević, M.S.; Simić, Z. Stark-Zeeman Line Shape Modeling for Magnetic White Dwarf and Tokamak Edge Plasmas: Common Challenges. *Atoms* **2017**, *5*, 36. [CrossRef]
62. Kieu, N.; Rosato, J.; Stamm, R.; Kovačević-Dojicinović, J.; Dimitrijević, M.S.; Popović, L.Č.; Simić, Z. A New Analysis of Stark and Zeeman Effects on Hydrogen Lnes in Magnetized DA White Dwarfs. *Atoms* **2017**, *5*, 44. [CrossRef]
63. Vogt, S.S.; Penrod, D.G. HIRES: A High Resolution Echelle Spectrometer for the Keck 10-Meter Telescope. In *Instrumentation for Ground-Based Optical Astronomy—Santa Cruz Summer Workshops in Astronomy and Astrophysics*; Robinson L.B., Ed.; Springer: New York, NY, USA, 1988.
64. W.M. Keck Observatory Archive (KOA). Extracted HIRES Spectra for the White Dwarf HG7-85, KOA-ID: HI.20121028.35667. Available online: https://koa.ipac.caltech.edu (accessed on 5 May 2018).
65. Zuckerman, B.; Xu, S.; Klein, B.; Jura, A. Thy Hyades cluster: Identification of a planetary system and ascaping white dwarfs. *Astrophys. J.* **2013**, *770*, 140–141. [CrossRef]
66. Keller, J. *Nonlinear Fitting n-Dimensional Data with Arbitrary Functions*; Mathworks File Exchange: Natick, MA, USA, 2008. Available online: https://www.mathworks.com/matlabcentral/fileexchange/20540-nonlinear-fitting-n-dimensional-data-with-arbitrary-functions (accessed on 14 February 2018).

67. Parigger, C.G.; Woods, A.C.; Witte, M.J.; Swafford, L.D.; Surmick, D.M. Measurement and Analysis of Atomic Hydrogen and Diatomic Molecular AlO, C_2, CN, and TiO Spectra Following Laser-induced Optical Breakdown. *J. Vis. Exp.* **2014**, *84*, e51250. [CrossRef]

68. Parigger, C.G.; Woods, A.C.; Surmick, D.M.; Gautam G.; Witte, M.J.; Hornkohl, J.O. Computation of diatomic molecular spectra for selected transitions of aluminum monoxide, cyanide, diatomic carbon, and titanium monoxide. *Spectrochim. Acta B At. Spectrosc.* **2015**, *107*, 132–138. [CrossRef]

69. Surmick, D.M.; Parigger, C.G. Empirical Formulae for Electron Density Diagnostics from H_α and H_β Line Profiles. *Int. Rev. At. Mol. Phys.* **2014**, *5*, 73–81, ISSN 2229-3159.

70. Büscher, S.; Wrubel, T.; Ferri, S.; Kunze, H.-J. The Stark width and shift of the hydrogen Hα line. *J. Phys. B At. Mol. Opt. Phys.* **2002**, *35*, 2889–2897. [CrossRef]

71. Kask, N.E.; Michurin, S.V. Broadening of the spectral lines of a buffer gas and target substance. *Quant. Electron.* **2012**, *42*, 1002–1007. [CrossRef]

72. Kask, N.E.; Lexsina, E.G.; Michurin, S.V.; Fedorov, G.M.; Chopornyak, D.B. Broadening and shift of the spectral lines of hydrogen atoms and silicon ions in laser plasma. *Quant. Electron.* **2015**, *45*, 527–532. [CrossRef]

73. Witkovsky, V. CharFunTool: The Characteristic Functions Toolbox. Matlab® File Exchange: Script cfX_PDF.m for Exponential Distributions in the CF_Repository. Available online: https://www.mathworks.com/matlabcentral/fileexchange/64400-charfuntool--the-characteristic-functions-toolbox (accessed on 19 March 2018).

74. Demura, A.V. Physical Models of Plasma Microfield. *Int. J. Spectrosc.* **2010**, *2010*, 671073. [CrossRef]

75. Griem, H.R. Stark Broadening of Higher Hydrogen and Hydrogen-like lines by electrons and ions. *Astrophys. J.* **1960**, *132*, 883–893. [CrossRef]

76. Griem, H.R.; Kolb, A.C.; Shen, K.Y. Stark Profile Calculations for the Hβ Line of Hydrogen. *Astrophys. J.* **1962**, *135*, 272–276. [CrossRef]

77. Sapar, A.; Poolamäe, R.; Sapar, L. High-precision approximation expressions for line profiules of hydrogenic particles. *Baltic Astron.* **2006**, *15*, 435–447.

78. Sapar, A.; Poolamäe, R. Revised line profile function for hydrogenic species. *Open Astron.* **2012**, *21*, 243–254. [CrossRef]

Review

Mini-Review of Intra-Stark X-ray Spectroscopy of Relativistic Laser–Plasma Interactions

Elisabeth Dalimier [1,*], Tatiana A. Pikuz [2,3] and Paulo Angelo [1]

[1] LULI—Sorbonne Université-Campus Pierre et Marie Curie, CNRS, Ecole Polytechnique, CEA: Université Paris-Saclay, CEDEX 05, F-75252 Paris, France; paulo.angelo@upmc.fr

[2] Joint Institute for High Temperatures, Russian Academy of Sciences, 125412 Moscow, Russia; pikuz.tatiana@gmail.com

[3] Open and Trans-Disciplinary Research Initiatives, Osaka University, 2-1, Yamadaoka, Suita, Osaka 565-0871, Japan

[*] Correspondence: elisabeth.dalimier@upmc.fr

Received: 10 July 2018; Accepted: 6 August 2018; Published: 16 August 2018

Abstract: Intra-Stark spectroscopy (ISS) is the spectroscopy within the quasi-static Stark profile of a spectral line. The present paper reviews the X-ray ISS-based studies recently advanced for the diagnostics of the relativistic laser–plasma interactions. By improving experiments performed on the Vulcan Petawatt (PW) laser facility at the Rutherford Appleton Laboratory (RAL), the simultaneous production of the Langmuir waves and of the ion acoustic turbulence at the surface of the relativistic critical density gave the first probe by ISS of the parametric decay instability (PDI) predicted by PIC simulations. The reliable reproducibility of the experimental signatures of PDI—i.e., the Langmuir-wave-induced dips—allowed measurements of the fields of the Langmuir and ion acoustic waves. The parallel theoretical study based on a rigorous condition of the dynamic resonance depending on the relative values of the ion acoustic and the Langmuir fields could explain the disappearance of the Langmuir dips as the Langmuir wave field increases. The ISS used for the diagnostic of the PDI process in relativistic laser–plasma interactions has reinforced the reliability of the spectral line shape while allowing for all broadening mechanisms. The results can be used for a better understanding of intense laser–plasma interactions and for laboratory modelling of physical processes in astrophysical objects.

Keywords: intra-Stark spectroscopy; relativistic laser–plasma interaction; parametric decay instability; X-ray spectral line profiles

1. Introduction

Novel X-ray spectral line shapes diagnostics in ultra-high intensity laser–plasmas have been constructed in recent years, uncovering a large amount information on the relativistic interactions characterizing the plasmas. This review is dedicated to the state-of-the-art of the improved spectroscopic methods adjusted for the analysis of the experimental results from Vulcan Petawatt (PW) laser facility at the Rutherford Appleton Laboratory (RAL) [1]. These methods are based on the intra-Stark spectroscopy (ISS) which is spectroscopy within the quasi-static Stark profile of a spectral line when radiating ions are subjected simultaneously to a quasi-static field F and to a quasi-monochromatic electric field $E(t)$ at the characteristic frequency ω. Due to non-linear phenomena, some dips occur at certain locations of the quasi-static Stark profile. These dips correspond to a non-linear dynamic resonance between the Stark splitting ω_F of the spectral line and the frequency ω of the electric field or its harmonics. Details on the theory of ISS can be found in [2–5].

In the plasmas of interest, the quasi-monochromatic electric field $E(t)$ represents a Langmuir wave. The corresponding structures in the spectral line profiles are the Langmuir-wave-induced dips (L-dips) and they were first exhibited in the gas-liner pinch experiments performed by Kunze's group [4] providing a passive spectroscopic method for measuring with a high accuracy the electron density in plasmas and the amplitude of the Langmuir waves produced by nonlinear physical processes. Then experimental studies of L-dips were implemented in laser-produced plasmas obtained first with moderate laser intensities $(1-2) \times 10^{14}$ W/cm^2 [6–10]. It was shown that the ISS X-ray spectroscopy of these plasmas helped to reveal the physics of the laser–plasma interaction, leading to the generation of the Langmuir field. Later, the experiments were performed using a femtosecond laser driven cluster-based plasma, the laser intensity $(0.4-3) \times 10^{18}$ W/cm^2 being at the threshold for relativistic laser–plasma interaction [11]. In these experiments the L-dips observed were caused by Langmuir waves resulting from the two-plasmon decay instability. The present review is totally dedicated to experimental studies performed with incident laser intensities as high as 10^{20}–10^{21} Wcm^{-2} and generating strongly relativistic laser–plasma interactions [12,13]. It advances the significant improvements developed at RAL, using the Optical Parametric Chirped Pulse Amplification technology (OPCPA) for the laser pulse generation and a plasma mirror for increasing the laser contrast and as a consequence diminishing the importance of the laser pre-plasma. High-resolution spectroscopy using a Focusing Spectrometer with Spatial Resolution (FSSR) allowed the analysis of SiXIV Ly-beta, Ly-gamma, and AlXIII Ly-beta lines. The reproducibility of the interpretation of spectroscopic results obtained from different shots of about the same laser intensity allowed the demonstration of the simultaneous production of the *Langmuir high frequency waves* and, for the first time, the *ion acoustic low frequency waves*, in high-density plasmas. Both kinds of waves have been predicted and attributed to the development of the non-linear physical process Parametric Decay Instability (PDI) developing at the surface of the *relativistic* critical density [13,14].

After an in-depth description of the experiments devoted to relativistic laser–plasmas at RAL in Section 2, the theory of ISS for those plasmas is reviewed in Section 3 including an overview of the theory of Langmuir-wave-caused dips for plasmas and two recent thorough theoretical studies: a robust computational method for fast calculations of line shapes affected by a low-frequency electrostatic plasma turbulence (i.e., the ion acoustic wave) and a rigorous analysis of the dynamic resonance condition involved in the ISS spectroscopy. The interpretation of the experimental profiles with the theoretical profiles is provided in Section 4. The support of the results with PIC simulations is detailed in Section 5. Finally, the conclusion (Section 6) enhances the first experimental discovery of the PDI non-linear process in relativistic laser–plasma interactions.

2. X-ray Spectra Measurements in Relativistic Laser–Plasma Interactions

The experiments were performed at Vulcan Petawatt (PW) facility at the Rutherford Appleton Laboratory [12,13], which provides a beam using optical Parametric, Chirped Pulse Amplification (OPCPA) technology with a central wavelength of 1054 nm and a pulse Full-Width-Half-Maximum (FWHM) duration, which could be up to 1500 fs. The OPCPA approach associated with a plasma mirror enables an amplified spontaneous emission to the peak-intensity contrast ratio exceeding 10^{-11} several nanoseconds before the peak of the laser pulse [15–17]. The laser pulse was focused with an f/3 off-axis parabola. At the best focus approximately 300 J on the target was contained within a 7 μm (FWHM) diameter spot providing a maximum intensity of 1.4×10^{21} W/cm^2. The horizontally polarised laser beam was incident on target at 45° from the target surface normal, as shown schematically in Figure 1a.

The X-ray emission of the plasma was registered by means of a high luminosity focusing spectrometer, with a spatial resolution (FSSR) [18], at the directions close to the normal to the target surface, from the rear side of the target. To obtain the spectra with a high spectral resolution $(\lambda/\delta\lambda \sim 3000)$ in a rather broad wavelength window the FSSR was equipped with quartz/mica spherically bent crystals [19,20] settled to record K-shell emission of Rydberg H-like spectral lines of multi charged SiXIV and AlXIII ions. For reducing the level of noise caused by the background fogging and crystal fluorescence, a pair of 0.5 T neodymium–iron–boron permanent magnets (not reproduced in Figure 1a), which formed a slit of 10 μm wide, was placed in front of the crystal. The spectra were recorded by means of Fujifilm TR Image Plate (IP) detectors protected against the exposure to the visible light by two layers of 1 μm-thick polypropylene $(C_3H_6)_n$ with 0.2 μm Al coating. Additionally, to prevent saturation of the detectors, the Mylar $(C_{10}H_8O_4)$ filter of 5 μm thickness was placed at the magnet entrance. Moreover, some of pure Si and Al targets were coated with or buried into thin plastic (CH) in order to keep the plasma of the main layer at solid densities regardless of the impact of the laser pre-pulse.

Figure 1b shows as examples the experimental spectra of Si XIV Ly$_\beta$ and Ly$_\gamma$ lines in two different shots with duration of 600 fs and laser intensities at the surface of the target theoretically estimated as 1.01×10^{21} W/cm^2 and 0.24×10^{21} W/cm^2, respectively. The positions of the dips/depressions in the profiles are marked in insets by vertical lines separated either by $2\lambda_{pe}$ or $4\lambda_{pe}$, where $\lambda_{pe} = [\lambda_0^2/(2\pi c)]\omega_{pe}$ (λ_0 is the unperturbed wavelength of the corresponding line and ω_{pe} the plasma electron frequency).

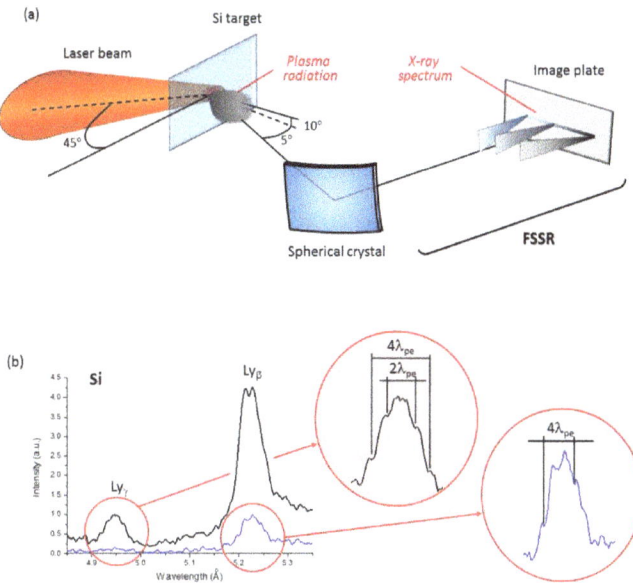

Figure 1. Schematic of experimental setup and typical X-ray spectra in the range of 0.485–0.535 nm. Experimental setup (**a**) and profiles (**b**) of Si XIV spectral lines, obtained in a single laser shot with initial laser intensity at the surface of the target estimated as 1.01×10^{21} W/cm^2 (black trace) and 0.24×10^{21} W/cm^2 (blue trace). In the insets, positions of the dips/depressions in the profiles are marked by vertical lines separated either by $2\lambda_{pe}$ or $4\lambda_{pe}$, where $\lambda_{pe} = [\lambda_0^2/(2\pi c)]\omega_{pe}$ (λ_0 is the unperturbed wavelength of the corresponding line and ω_{pe} the plasma electron frequency) [12].

3. Theory of Intra Stark Spectroscopy ISS for Relativistic Plasmas

3.1. The Langmuir-Wave-Caused Dips for Plasmas Dominated by a Low Frequency Electrostatic Turbulence LET

The Langmuir-wave-caused dips (L-dips) are the signatures of non-linear processes with generation of Langmuir high frequency waves. They result from a multifrequency resonance between the Stark splitting $\omega_F = 3n\hbar F/(2Z_r m_e e)$ of the energy levels (given here for hydrogen lines), caused by the quasi-static field F in the plasma, and the frequency ω_L of the Langmuir waves, which practically coincides with the plasma electron frequency $\omega_{pe}(Ne)$: $\omega_F = s\omega_{pe}(N_e)$, $s = 1, 2, \ldots$ Here n and Z_r are the principal quantum number and the nuclear charge of the radiating atom/ion (radiator) respectively,

Ne stands for the electronic density and *s* for the number of quanta of the field involved in the resonance, $\hbar$ is the Plank constant, *e* and m_e are the electron charge and mass [2]. The quasi-static field *F* represents the low-frequency part of the ion micro-field and the low Frequency Electrostatic Turbulence (LET), if the latter is developed in the plasma. This will be the case in relativistic plasmas as predicted by PIC simulations [12]. The physical mechanism producing simultaneously the Langmuir waves and the LET (specifically, the ion acoustic turbulence) out of the laser field is the Parametric Decay Instability (PDI). PDI is a nonlinear process, in which an electromagnetic wave decays into a Langmuir wave and an ion acoustic wave at the surface of the critical density N_c. It is important to remark that due to the broad distribution of the low frequency field *F*, there is always a fraction of radiators for which the resonance condition is satisfied. Even though the electric field of the Langmuir wave is considered to be monochromatic ($E_0 \, cos\omega_{pe} \, t$), it produces a multi-frequency resonance as has been shown in paper [5].

The resonance condition $\omega_F = s \, \omega_{pe}(N_e)$, translates into specific locations of L-dips in spectral line profiles depending on N_e. In the following two specific cases are discussed for the Ly-lines.

In the case where the quasi-static field *F* is dominated by the LET, the distance of an L-dip from the unperturbed wavelength λ_0 is given by $\Delta\lambda_{dip}(q, N_e) = -\left[\lambda_0^2/(2\pi c)\right]qs\omega_{pe}(N_e)$. Here $q = n_1 - n_2$ is the electric quantum number expressed via the parabolic quantum numbers n_1 and n_2: $q = 0, \pm 1, \pm 2, \ldots, \pm(n-1)$. It labels Stark components of Ly-lines. For a pair of Stark components, corresponding to *q* and $-q$, there could be multi-quantum resonances ($s = 1, 2 \ldots$) leading to different pairs of L-dips located symmetrically in the red and blue parts of the profile: $\Delta\lambda_{dip}(N_e) = \pm \left[\lambda_0^2/(2\pi c)\right]qs\omega_{pe}(N_e)$.

If the quasi-static field *F* is dominated by the ion micro-field, then the above formula for $\Delta\lambda_{dip}$ would hold only for relatively low electron densities. For relatively high electron densities, the spatial non-uniformity of the ion micro-field has to be taken into account leading to the result [2,21]:

$$\Delta\lambda_{dip}(q, N_e) = -\frac{\lambda_0^2}{2\pi c}\left[qs\omega_{pe} + \frac{2(s\omega_{pe})^3}{27n^3 Z_r Z_p \omega_{at}}\right]^{\frac{1}{2}}\left[n^2\left(n^2 - 6q^2 - 1\right) + 12n^2 q^2\right] \tag{1}$$

where Z_p is the charge of the perturbing ions, Z_r the charge of the radiating ions and $\omega_{at} = m_e e^4/\hbar^3 = 4.14 \cdot 10^{16} s^{-1}$ the atomic unit of frequency. The first, primary term in braces reflects the dipole interaction with the ion micro-field. The second, smaller term in braces takes into account—via the quadrupole interaction—a spatial non-uniformity of the ion micro-field. It results in the shift of the midpoint between the pair of L-dips, corresponding to *q* and $-q$, with respect to the unperturbed wavelength. Thus, there is a signature of the pre-dominance of the LET when the symmetrical location of the L-dips in the red and blue parts of the experimental profile is observed.

In any case (whether the field *F* is dominated by the ion micro-field or by a LET), the separation of the two L-dips, corresponding to *q* and $-q$, is:

$$\Delta\lambda_{dip}(N_e) = \pm \left[\lambda_0^2/(2\pi c)\right]q\omega_{pe}(N_e) \tag{2}$$

thus enabling measurement of N_e. It is important to emphasize that this passive spectroscopic method for measuring N_e is just as accurate as the active spectroscopic method using the Thompson scattering, [4,21]. From the experimental ISS results analysis, the simultaneous signatures of the Langmuir waves and the LET confirmed that the PDI nonlinear process actually took place at the critical density N_c.

The half-width of the L-dip (i.e., the separation between the dip and the nearest "bump"), is controlled by the amplitude E_0 of the Langmuir wave [2]:

$$\delta\lambda_{1/2} \approx \left(\frac{3}{2}\right)^{\frac{1}{2}} \frac{\lambda_0^2 n^2 E_0}{8\pi m_e e c Z_r} \tag{3}$$

Thus, by measuring the experimental half-width of L-dips, one can determine the amplitude E_0 of the Langmuir wave.

Before analyzing the experimental spectral lines, we should note the following. The laser frequency ω corresponds to the wavelength λ of approximately 1054 nm. At lower, non-relativistic laser intensities, the theoretical electron density N_e determined from the equation $\omega = \omega_{pe}$, where $\omega_{pe} = (4\pi e^2 N_e / m_e)^{\frac{1}{2}}$, would be 1.0×10^{21} cm^{-3}. However, at the laser intensities $\sim 10^{21}$ W/cm^2, i.e., those corresponding to the experiment, due to relativistic effects, the "relativistic" critical electron density N_{cr} becomes higher than the critical density N_c [22–24]. For the linear polarized laser radiation, it becomes [25]:

$$N_{cr} = \frac{\left(\frac{\pi a}{4}\right) m_e \omega^2}{4\pi e^2}, a = \lambda(\mu m)\left[\frac{I\left(\frac{W}{cm^2}\right)}{1.37 \times 10^{18}}\right]^{\frac{1}{2}} \tag{4}$$

The ability of the ultra-intense laser radiation to penetrate into regions of the density higher than N_c—i.e., a relativistically-induced transparency regime—has been previously demonstrated by PIC simulations [26]. In the present paper, we not only took into account the effect of the relativistically induced transparency (details of which can be found in paper [27]) on the resonance condition, but also confirmed that the L-dips originate from the region of the relativistic critical density (which is greater than the non-relativistic critical density), as shown in Section 4.

3.2. Rigorous Condition of the Dynamic Resonance

The resonance condition for the formation of the dips $\omega_{stark}(F_{res}) = s\omega_{pe}(N_e)$ with $\omega_{stark}(F_{res}) = 3n\hbar F_{res}/(2Z_r m_e e)$ (deeply studied for hydrogen lines in Section 3.1) is only valid for $E_0/F_{res} \ll 1$, this condition allowing the determination of the resonant value of the quasi-static field F_{res} independent of E_0.

The rigorous resonant condition valid for any value of the ratio E_0/F_{res} reads:

$$\omega_{stark}(F_{res}) g(\varepsilon) = s\omega_{pe}(N_e), g(\varepsilon) = (1 + \varepsilon^2)^{1/2} \ EllipticE[\varepsilon/(1 + \varepsilon^2)^{1/2}], \varepsilon = E_o/F_{res}, \tag{5}$$

where $\omega_{stark}(F_{res}) = 3n\hbar F_{res}/(2Z_r m_e e)$ for hydrogen lines and *EllipticE*[...] is the complete elliptic integral of the second kind [2] (pp. 49, 28; Appendix G). When the ratio $E_0/F_{res} > 0.5$, L-dips cannot exist and instead, in the profile, there could appear a sequence of satellites at the resonance frequencies [28] (Appendix L). It is important to emphasize that this resonance condition involving now both E_0 and F_{res} (5) is relevant for some shots during the experimental campaigns, explaining the disappearance of the Langmuir dips.

Several notes should be made. First, in our experimental conditions, the low-frequency electric field in the plasma is dominated by the LET, rather than by the ion microfield. Second, the resonant value of the low-frequency electric field F_{res}, defined by Equation (5), is much greater than the characteristic value F_{ion} of the ion microfield. Therefore, the field of the value F_{res} can be considered quasi-static even if much smaller fields $\sim F_{ion}$ would not be quasi-static. Third, we did not assume the entire ion microfield to be quasi-static.

3.3. Robust Computation of Line Shapes Affected by the LET

For a detailed quantitative analysis of the ISS results from relativistic laser–plasma, we calculated the theoretical profiles as follows. In this computation, the total quasi-static field F is the vector sum of two contributions: $F = F_t + F_i$. The first contribution F_t is the field of the LET, while the second contribution F_i is the quasi-static part of the ion micro-field. We employed the results from papers [29–31] to calculate the distributions of the total quasi-static field $F = F_t + F_i$. Specifically, we calculated the distribution of the total quasi-static field F in the form of the convolution of the APEX (Adjustable-Parameter EXponential approximation) distribution of F_i [30] with the Rayleigh-type distribution of F_t [29]. We note that for the case where the characteristic value of the LET is much greater than the characteristic value of the ion micro-field, the analytical results from paper [31] provide a robust way to calculate the distribution of the total quasi-static field without calculating the convolution. We also took into account the broadening by the electron micro-field, by the dynamical part of the ion micro-field, the Doppler and the instrumental broadenings (~0.1 mÅ), as well as the theoretically expected asymmetry of the profiles [32,33]. For calculating the details of the spectral line shape in the regions of L-dips we employed the analytical solution from [5] for the wave functions of the quasi-energy states, the latter being caused simultaneously by all harmonics of the total electric field $E(t) = F + E_0 \cos(\omega_{pe}t)$ (vectors F and E_0 are not collinear).

In the present paper, all theoretical profiles obtained with the above robust computations will be compared to the results performed using the code FLYCHK [34]. This code, that does not take into account the Stark broadening by the LET and the presence of the L-dips, leads, when comparing with the experimental results, to very high electronic densities $1\text{--}3 \ 10^{23}$ cm^{-3} inconsistent with the relativistic critical electron density $N_{cr} = 2\text{--}3 \ 10^{22}$ cm^{-3}. The variation of the FLYCHK code used takes into account both Doppler and instruments broadenings.

4. Analysis of the Experimental Spectra and Comparison with Spectral Line Shapes Codes Simulations

4.1. First Discovery of the Ion Acoustic Turbulence in Relativistic Laser–Plasmas Using X-ray Spectroscopy

Let us consider the experimental profiles of Si XIV Ly$_\gamma$, produced in an interaction with a single laser pulse with initial laser intensity at the surface of the target estimated as 1.01×10^{21} W/cm^2 (black trace in Figure 1b), and Si XIV Ly$_\beta$, produced in an interaction with a single laser pulse with initial laser intensity at the surface of the target estimated as 0.24×10^{21} W/cm^2 (blue trace in Figure 1b) [12]. For these profiles the peak-intensity contrast ratio was 10^9. We mention that L-dips were not observed in the experimental profile of Si XIV Ly$_\beta$ line in the black trace Figure 1b. It seems that a process of self-absorption of radiation prevented L-dips from being visible in this line. We note that without the self-absorption the Ly$_\beta$ line would show a relatively deep central minimum (related not to the the L-dips, but to the structure of the Stark splitting of this line) and shallow L-dips. The self-absorption completely washes out the L-dips but does not completely washes out the central minimum: it just makes it shallow instead of being relatively deep. As for the Ly$_\gamma$ line in the black trace Figure 1b, at the possible locations of the L-dips, the experimental profile already merged into the noise.

The experimental black trace profile Si XIV Ly$_\gamma$ shows a pair of two L-dips nearest to the line center ($q = \pm 1$, $s = 1$) separated from each other by 28 mÅ yielding an electron density of $N_e = 3.6 \times 10^{22}$ cm^{-3}. In addition to this, a pair of L-dips, separated from each other by 56 mÅ, were also observed, representing a superposition of two pairs of the L-dips: $q = \pm 2$, $s = 1$ and $q = \pm 1$, $s = 2$. From the Equation (2) this yields the same $N_e = 3.6 \times 10^{22}$ cm^{-3}, thus reinforcing the interpretation of the experimental dips as the L-dips—caused by the resonant interaction of the Langmuir waves, developed at the surface of the relativistic critical density, with the quasi-static electric field. Let us remark that at the initial laser intensity at the surface of the target estimated as 1.01×10^{21} W/cm^2, the relativistic critical density would be $N_{cr} = 2.3 \times 10^{22}$ cm^{-3} according to Equation (4). However, due to several physical effects [26,27] (self-focusing of the laser beam, Raman and Brillouin back scattering), the actual intensity of the transverse electromagnetic wave in the plasma can be significantly higher than the intensity of the incident laser radiation at the surface of the target. In the present experiment for the relativistic critical density to be approximately equal to the density $N_e = 3.6 \times 10^{22}$ cm^{-3} deduced by the spectroscopic analysis, it would require the enhancement of the initially estimated intensity of the transverse electromagnetic wave at the surface of the target due to the above physical effects just by a factor of two. In both cases, L-dips separated by 28 mÅ and for the L-dips separated by 56 mÅ, the mid-point between the two dips in the pair practically coincides with the unperturbed wavelength λ_0. This is a strong indication that the quasi-static field F was dominated by the LET. If the LET would be absent, then according to Equation (1), the mid-point of the pair of the L-dips separated by 28 mÅ should have been shifted by 5.8 mÅ to the red with respect to λ_0 and the mid-point of the pair of the L-dips separated by 56 mÅ would have been similarly shifted by 10.7 mÅ to the red with respect to λ_0, these shifts being due to the spatial non-uniformity of the ion micro-field reflected by the second term in Equation (1). Another strong indication of the presence of the LET observed from the above result comes from the analysis of the broadening of this spectral line that was performed using code FLYCHK [34]. This code, which does not take into account the Stark broadening by the LET (and the presence of L-dips), yielded $N_e = 0.9 \times 10^{23}$ cm^{-3}, i.e., almost three times higher than the actual $N_e = 3.6 \times 10^{22}$ cm^{-3} for the best fit to the experimental profile as shown in Figure 2a. This result is not reasonable.

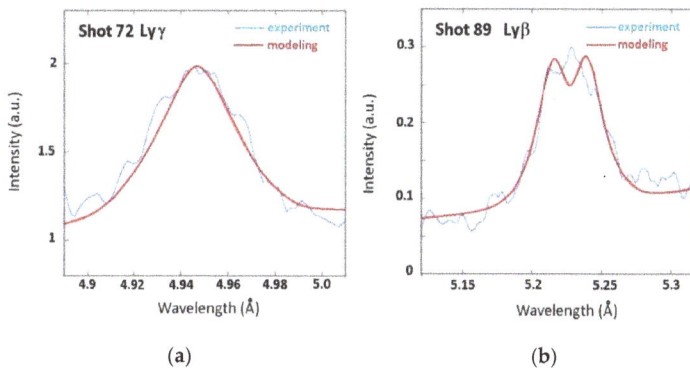

Figure 2. Experimental spectra and their comparisons with FLYCHK modeling. (**a**) Comparison of the experimental profile of Si XIV Lyγ line in shot with initial laser intensity at the surface of the target estimated as 1.01×10^{21} W/cm^2 with a simulation performed using a variation of the code FLYCHK for calculating the Stark broadening, then adding both Doppler and instrument broadening, and if necessary, opacity. The L-dips phenomenon and the spectral line broadening by LET were not included in the FLYCHK. The plasma parameters for the best fit are $N_e = 0.9 \times 10^{23}$ cm^{-3} and $T_e = 500$ eV. (**b**) The same as in (**a**) but for Si XIV Lyβ line produced in a single laser shot with initial laser intensity at the surface of the target estimated as 0.24×10^{21} W/cm^2; $N_e = 3 \times 10^{23}$ cm^{-3} and $T_e = 500$ eV [12].

The analysis of the experimental profile of Si XIV Ly$_\beta$ with the initial laser intensity at the surface of the target estimated as 0.24×10^{21} W/cm^2 (blue trace in Figure 1b), shows a situation similar to Si XIV Ly$_\gamma$. There is a pair of the L-dips separated from each other by 43 mA. The electron density deduced from the separation within the pair of these L-dips, is $N_e = 1.74 \times 10^{22}$ cm^{-3} (assuming $|q|s = 2$ in Equation (2)). This pair of dips corresponds either to the Stark component $q = 1$ for a two-quantum resonance $s = 2$ or to the Stark component $q = 2$ for the one-quantum resonance $s = 1$. The superposition of two different dips at the same location results in a L-super-dip with a significantly enhanced visibility. The location of the would-be L-dips, corresponding to $|q|s = 1$, is too close to the central, most intense part of the experimental line profile so that they are not observed either due to a self-absorption in the most intense part of the profile, or because the relatively small values of the field F, corresponding to the central part of the profile, are not quasi-static. The mid-point between the two dips in the pair practically coincides with the unperturbed wavelength λ_0. This is again a strong indication that the quasi-static field F was dominated by the LET. If the LET would be absent, then according to Equation (1), the mid-point of this pair of L-dips should have been shifted by 6.1 mÅ to the red with respect to λ_0. Another strong indication of the presence of the LET in the profile of Si XIV Ly$_\beta$ comes from the analysis performed using code FLYCHK [34] (Figure 2b). An electron density of $N_e = 3 \times 10^{23}$ cm^{-3}, was obtained, i.e., 17 times higher than the experimentally verified $N_e = 1.74 \times 10^{22}$ cm^{-3}. This is yet another strong indication of an additional Stark broadening by the LET (not accounted for by FLYCHK). At the initial laser intensity at the surface of the target estimated as 0.24×10^{21} W/cm^2, the relativistic critical density would be $N_{cr} = 1.1 \times 10^{22}$ cm^{-3} according to Equation (4). However, again due to the self-focusing, as well as Raman and Brillouin backscattering, the actual intensity of the transverse electromagnetic wave in the plasma can be significantly higher. In the present case, for the relativistic critical density to be approximately equal to the density $N_e = 1.74 \times 10^{22}$ cm^{-3} deduced by the spectroscopic analysis, again it would require the enhancement of the intensity of the transverse electromagnetic wave due to the above physical effects just by a factor of two.

From both analysis of the Si XIV Ly$_\gamma$ and Si XIV Ly$_\beta$ profiles, it can be confirmed that the LET developed simultaneously with the Langmuir waves at the relativistic critical density surface and thus it is most likely to be an ion acoustic turbulence. The most probable and the best studied mechanism for developing Langmuir waves at the surface of the relativistic critical density is a parametric decay, which is a nonlinear process where the pump wave (t_1) excites both the Langmuir wave (l) and an ion-acoustic wave (s): $t_1 \rightarrow l + s$.

4.2. Robust Computations for the Analysis of the Experimental Si XIV Ly$_\gamma$ and Si XIV Ly$_\beta$ Profiles

The robust computations including L-dips and the spectral line broadening by LET (Section 3.3) performed for the analysis of the Si XIV Ly$_\gamma$ and Si XIV Ly$_\beta$ profiles are given in Figure 3a,b. The comparisons with the experimental spectra is fruitful as confirmed in the following.

From the experimental profile of Si XIV Ly$_\gamma$ (Figure 3a), it follows that the root-mean-square value of F_t was $F_{t,rms} = 2.1$ GV/cm. For comparison, the characteristic ion microfield $F_{i,typ} = 2.603\, eZ^{1/3}N_e^{2/3}$ was 1.0 GV/cm. From the half width of the experimental L-dips, by using Equation (3), we found the amplitude of the Langmuir wave to be $E_0 = 0.6$ GV/cm. The resonant value of the quasi-static field F_{res}, determined by the condition of the resonance between the separation of the Stark sublevels and the plasma frequency $3n\hbar F_{res}/(2Z_r m_e e) = \omega_{pe}$, was 3.1 GV/cm, so that the validity condition for the existence of L-dips $E_0 < F_{res}$ (Section 3.2) was satisfied.

From the experimental profile of Si XIV Ly$_\beta$ (Figure 3b), it follows that the root-mean-square value of F_t was $F_{t,rms} = 3.9$ GV/cm. For comparison, the characteristic ion microfield was $F_{i,typ} = 0.6$ GV/cm. From the halfwidth of the experimental L-dips, by using Equation (3), we found the amplitude of the Langmuir wave to be $E_0 = 1.0$ GV/cm. The resonant value of the quasi-static field was $F_{res} = 5.8$ GV/cm, so that the validity condition for the existence of L-dips $E_0 < F_{res}$ was again satisfied.

The good agreement between experimental spectra and simulations reinforces the discovery of the simultaneous production of LET with the Langmuir waves. We note that the electron densities involved turned out to be much lower than the densities deduced using FLYCHK simulations, which ignored the LET and the L-dips. The densities used for the robust computations are in perfect agreement with the densities deduced experimentally from the dips separations.

We also performed experiments where the spectrometer viewed the laser-irradiated front surface of the target. As an example, Figure 4 shows the experimental spectrum of Al XIII Ly$_\beta$ line (4 µm Al foil coated by 0.45 µm CH), which was obtained in a single laser shot with duration of 0.9 ps and the laser intensity at the surface of the target theoretically estimated as 6.7×10^{20} W/cm^2. The peak-intensity contrast ratio was 10^9 as for the experimental spectra Si XIV Ly$_\beta$ and Ly$_\gamma$. This spectrum exhibits two pairs of L-dips: one pair—at ± 16.8 mA from the line center, another pair—at ± 33.6 mA from the line center. The two dip pairs (two pairs of dips) are symmetrical with respect to the line center, confirming the production of LET with the Langmuir waves. Also, the density used for the robust computation is in perfect agreement with the density deduced experimentally from the dips separations.

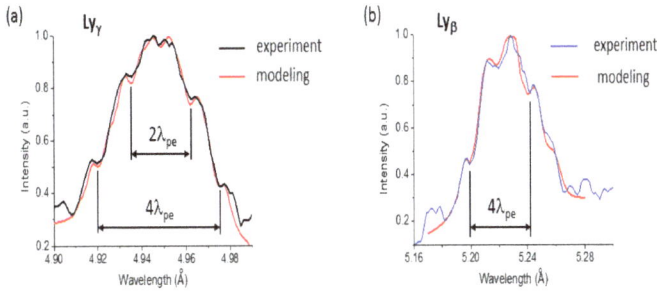

Figure 3. Experimental spectra and their comparisons with a different Stark broadening code including L-dips and the spectral line broadening by LET. (**a**) Experimental spectra of Si XIV Ly_γ line in a single laser shot with initial laser intensity at the surface of the target estimated as 1.01×10^{21} W/cm^2 initial laser intensity at the surface of the target. The positions of the dips/depressions in the profiles are marked by vertical lines separated either by $2\lambda_{pe}$ or $4\lambda_{pe}$, where $\lambda_{pe} = [\lambda_0^2/(2\pi c)]\omega_{pe}$ (λ_0 is the unperturbed wavelength of the corresponding line). Also shown is a theoretical profile at $N_e = 3.6 \times 10^{22}$ cm^{-3} allowing, in particular, for a low-frequency electrostatic turbulence. (**b**) Same but for Si XIV Ly_β line in a single laser shot with initial laser intensity at the surface of the target estimated as 0.24×10^{21} W/cm^2 initial laser intensity at the surface of the target. The theoretical profile is shown at $N_e = 1.74 \times 10^{22}$ cm^{-3} [12].

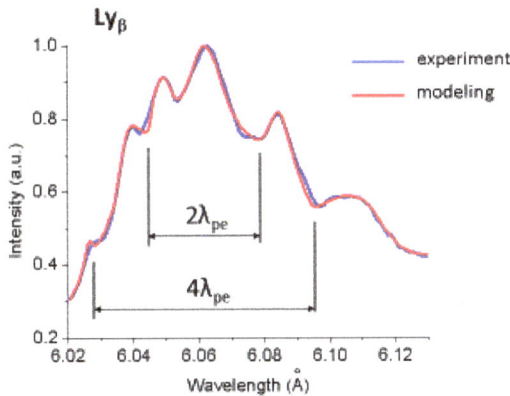

Figure 4. Experimental spectrum of Al XIII Ly_β and its comparison with the Stark broadening code including L-dips and the spectral line broadening by LET. The experimental spectrum was obtained in a single laser shot with initial laser intensity at the surface of the target estimated as 6.7×10^{20} W/cm^2. The positions of the L-dips in the profiles (the L-dips being very pronounced) are marked by vertical lines separated either by $2\lambda_{pe}$ or $4\lambda_{pe}$, where $\lambda_{pe} = [\lambda_0^2/(2\pi c)]\omega_{pe}$ (λ_0 is the unperturbed wavelength of the corresponding line). Also shown is a theoretical profile at $N_e = 2.35 \times 10^{22}$ cm^{-3} allowing, in particular, for L-dips and for a low-frequency electrostatic turbulence. This value of Ne is much lower than Ne obtained from fitting the same experimental spectrum by code FLYCHK that did not include the L-dips phenomenon and the spectral line broadening by LET, and therefore had significant discrepancies with the experimental profile at the locations of the L-dips [12].

4.3. In-Depth Study of ISS in the X-ray Range Emission from Relativistic Laser–Plasmas

Recently new experiments have been performed at RAL [17] allowing a peak-intensity contrast ratio exceeding 10^{-11} (instead of 10^{-9} for the experimental results discussed in the paper up to now) due to a plasma mirror allowing an amplified spontaneous emission. An in-depth study was then undertaken for the case Si XIV Ly$_\beta$ allowing a reliable reproducibility of the Langmuir-wave-induced dips at the same locations in the experimental profiles as well as of the deduced parameters (fields) of the Langmuir waves and ion acoustic turbulence in several individual 1 ps laser pulses and of the peak irradiance of 1 to 3 × 10^{20} W/cm^2.

Figure 5 shows the comparisons of the robust computations, allowing in particular, for the LET and L-dips, with the corresponding experimental profiles from shots A, B, C. The L-super-dip is observed twice in the experimental profiles: one in the blue part and the other in the red part. These L-super-dips are located practically symmetrically at the distance $\Delta\lambda_{dip}(N_e)$ = 24 mÅ from the unperturbed wavelength. According to Equation (3) with $|q|s$ = 2, this translates into the electron density N_e = 2.2 × 10^{22} cm^{-3} The theoretical profiles were calculated for this density 2.2 × 10^{22} cm^{-3} and the temperature T = 600, 550, and 600 eV for shots A, B, and C, respectively. The comparison demonstrates a good agreement between the theoretical and experimental profiles, and thus reinforces the good interpretation of these experimental profiles. Moreover, it reinforces that it was the PDI at the surface of the relativistic critical density that produced simultaneously the Langmuir waves and the ion acoustic turbulence in shots A, B, and C. The modeling of the experimental profiles A, B, C using the code FLYCHK [34] (Figure 6) confirmed once more the need to introduce the LET for the interpretation. This innovative code yielded T = 500 eV and N_e = 6 × 10^{23} cm^{-3}. This value of N_e is one and a half orders of magnitude higher than the electron density N_e = 2.2 × 10^{22} cm^{-3} deduced from the experimental L-dips.

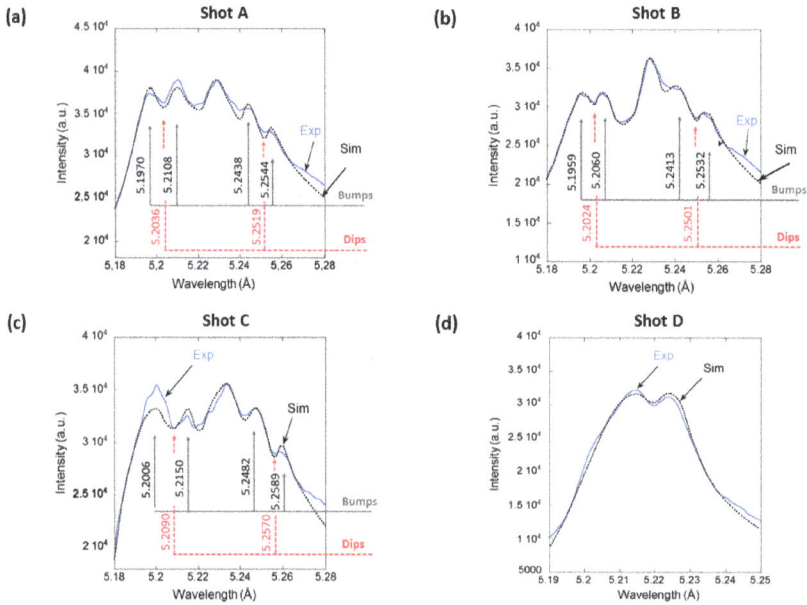

Figure 5. Comparison of the experimental profiles of the Si XIV Ly-beta line (solid line, blue in the online version, marked Exp) with the theoretical profiles (dotted line, black, marked Sim) allowing for the effects of the Langmuir waves, the LET, and all other broadening mechanisms (see the text). In the profiles from shots A, B, and C, there are clearly seen 'bump–dip–bump' structures (both in the red and blue parts of the profiles) typical for the L-dips phenomenon. The L-dips are not observed in shot D. The following parameters provided the best fit: (**a**) N_e = 2.2 × 10^{22} cm^{-3}, T = 600 eV, $F_{t,rms}$ = 4.8 GV/cm, E_0 = 0.7 GV/cm; (**b**) N_e = 2.2 × 10^{22} cm^{-3}, T = 550 eV, $F_{t,rms}$ = 4.4 GV/cm, E_0 = 0.5 GV/cm; (**c**) N_e = 2.2 × 10^{22} cm^{-3}, T = 600 eV, $F_{t,rms}$ = 4.9 GV/cm, E_0 = 0.6 GV/cm; (**d**) N_e = 6.6 × 10^{21} cm^{-3}, T = 550 eV, $F_{t,rms}$ = 2.0 GV/cm, E_0 = 2.0 GV/cm [13].

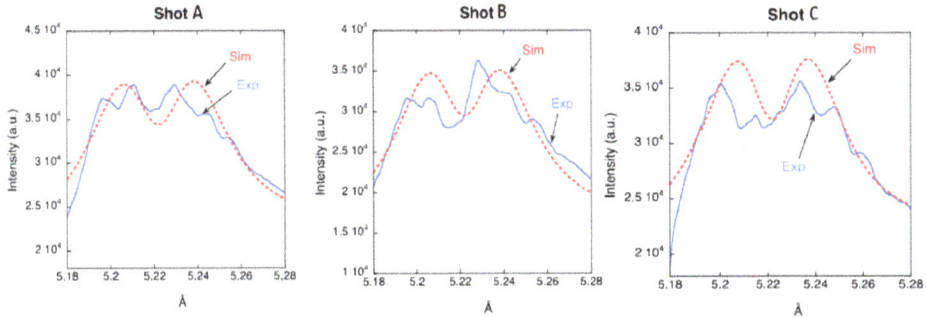

Figure 6. Experimental profiles of the Si XIV Ly-beta line in shots A, B, C (blue color in the online version) and their comparison with simulations using code FLYCHK (red color in the online version) at the electron density $N_e = 6 \times 10^{23}$ cm^{-3} and the temperature $T = 500$ eV [13].

The determination of the fields of the Langmuir waves and ion acoustic turbulence can be deduced from the experimental Si XIV Ly$_\beta$ profiles A, B, C (Figure 5) and their robust computations. The values of the amplitude of the Langmuir waves using (3) have been obtained: $E_0 = 0.7, 0.5$, and 0.6 GV/cm for shots A, B, and C, respectively. The resonant value of the quasi-static field F_{res} responsible for the formation of the L-dips, can be determined from the resonance condition $\omega_F = s\,\omega_{pe}(N_e)$. For shots A, B, and C, it yields: $F_{res} = 6.5$ GV/cm for $s = 2$ and $F_{res} = 3.25$ GV/cm for $s = 1$. These values of F_{res} are about 10 and 5 times higher than the Langmuir wave amplitude E_0, respectively. Thus, the condition $E_0 \ll F_{res}$, necessary for the formation of the L-dips, was fulfilled. These values for F_{res} are coherent with the root-mean-square field values of the LET introduced for the robust computations: $F_{t,rms} = 4.8, 4.4$, and 4.9 GV/cm for shots A, B, and C respectively. For comparison, the characteristic ion micro-field $F_{i,typ} = 2.603\ eZ1/3Ne2/3$ was 1.5 GV/cm.

Now we proceed to analyze the experimental profile of the Si XIV Ly$_\beta$ line in shot D (Figure 5). The experimental profile does not show bump–dip–bump structures—in distinction to shots A, B, and C. In shot D, the incident laser intensity was $I = 8.8 \times 10^{19}$ W/cm^2, significantly lower than in shots A, B, and C. The corresponding relativistic critical density is $N_{cr} = 6.6 \times 10^{21}$ cm^{-3}. The modeling of this spectra using the code FLYCHK yielded $N_e = 1.7 \times 10^{23}$ cm^{-3}, which is one and a half order of magnitude higher than the relativistic critical density and by an order of magnitude higher than the region of the electron density $N_e = 2.2 \times 10^{22}$ cm^{-3} from which the experimental profiles were emitted in shots A, B, and C is not reliable for the interpretation of shot D profile. The most probable interpretation of the experimental profile in shot D is the following.

In shot D the electron density was significantly lower than in shots A, B, and C. Therefore, the damping of the Langmuir waves was significantly lower, which could allow the Langmuir waves to reach a significantly higher amplitude. Figure 5d shows the comparison of the experimental profile from shot D with robust computing based on the code allowing for the LET and the Langmuir waves at $N_e = 6.6 \times 10^{21}$ cm^{-3}, $T = 550$ eV, $F_{t,rms} = 2.0$ GV/cm, $E_0 = 2.0$ GV/cm. It is seen that this theoretical profile is in a good agreement with the experimental profile and it does not exhibit bump–dip–bump structures. For high Langmuir wave amplitudes—i.e., when the ratio $E_0/F_{res} > 0.5$—the L-dips cannot form (see Section 3.3) and the resonant value of the quasi-static field is given by (5). For $N_e = 6.6 \times 10^{21}$ cm^{-3} and $E_0 = 2.0$ GV/cm, Equation (5) yields $F_{res} = 1.7$ GV/cm for $s = 1$ (so that $E_0/F_{res} = 1.2$) and $F_{res} = 3.5$ GV/cm for $s = 2$ (so that $E_0/F_{res} = 0.6$). Thus, both for the one-quantum resonance ($s = 1$) and for the two-quantum resonance ($s = 2$), we get $E_0/F_{res} > 0.5$, so that the L-dips were not able to form. We note that, while for shots A, B, and C the electron density could be deduced from the locations of the L-dips and from the robust computation, in shot D the only one possibility is by modeling the entire experimental profile using the code that allows for the interplay of the LET and the Langmuir waves. Up to now, it seems to be the first experimental profile diagnosed in relativistic laser–plasma interactions and explained with the simultaneous production of Langmuir waves and ion acoustic turbulence, but with no dips exhibited.

5. PIC Simulations for Relativistic Laser–Plasma Interactions

The results of this spectroscopic analysis have been also supported by PIC simulations. The simulations were performed using the modified 1D PIC code LPIC, in which a laser pulse of duration $t_L = 0.6$ ps and intensity $I_L = 2.7 \times 10^{21}$ W/cm^2 interacted with a Si target [35]. Laser pulse propagates along the x-axis and interacts with Si^{+14} inhomogeneous plasma layer with a linear density ramp over the length L = 6 μm and a target thickness of 2 μm, with a constant ion density of $N_i = 6 \times 10^{22}$ cm^{-3}. Angle of incidence was set up 45° at P polarization. The results of the simulations are shown in Figure 7.

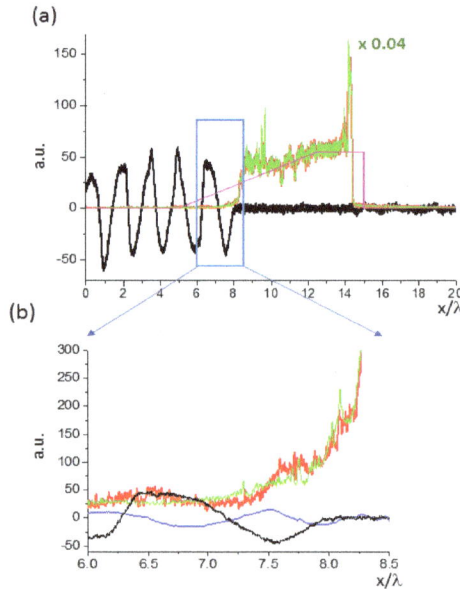

Figure 7. PIC simulations of the interaction of a 600 fs pulse of the intense $(2.7 \times 10^{21}$ W/cm$^2)$ linearly-polarized laser radiation with the Si^{14} plasma layer. The general view of the laser-target interaction at the instant $t = 320$ fs from the beginning of the interaction (**a**) and the zoom (**b**) on the area shown by the blue rectangle in (**a**). The coordinate x is in units of the laser wavelength λ. The initial scaled density profile is presented by the violet line. The scaled electron and ion densities are presented by the red and green lines. They are given in the units of $0.04ZN_{cr}$ in (**a**) and in the units of ZN_{cr} in (**b**); here $N_{cr} = \gamma N_c$, where N_c is the critical plasma density and γ is the relativistic factor. The calculated scaled transversal electric field $a_L = eE_L/(m_e\omega_L c)$ is shown by the black line and the longitudinal one (the Langmuir wave) $a_l = eE_l/(m_e\omega_L c)$—by the blue line [12].

At the peak of the laser pulse intensity $(t = 320$ fs$)$ it is seen that the scale of laser field decay is close to the scale of the plasma inhomogeneity. The longitudinal field of the Langmuir wave E_l appeared in the vicinity of the point where the density is about one-quarter of the relativistic critical density. The Langmuir wave exists up to a point slightly above the relativistic critical density $N_{cr} = 3.6 \times 10^{22}$ cm^{-3}. In Figure 7b, the region near the relativistic critical density is $6\lambda < x < 8.25\lambda$: the small modulation of ion density in this region shown by the green line is the manifestation of the ion acoustic wave. Similar processes of such parametric decays $(t \rightarrow l + l', s)$ were studied in the past [36], but for significantly lower laser intensities.

6. Conclusions

The evolution of the Intra-Stark X-ray Spectroscopy (ISS) in relativistic laser–plasma interactions during the last two years is presented here with a recent in-depth study of the simultaneous productions of Langmuir waves and of the ion acoustic turbulence at the surface of the relativistic critical density. We demonstrated, by improving the experiment at RAL, a reliable reproducibility of the spectroscopic signatures, i.e., the Langmuir dips of the Parametric Decay Instability (PDI) predicted by PIC simulations. The robust computations confirmed in details the experimental profiles, yielded the field parameters (Langmuir waves and ion acoustic turbulence), and demonstrated that the PDI process occurs at the relativistic critical density which is confirmed by the separation of the L-dips. The experiments revealed for the first time the situation where the PDI process could not lead to the appearance of L-dips when the Langmuir wave field is much larger than the LET field.

This study expands the Intra-Stark Spectroscopy (ISS) to relativistic laser–plasma situations and shows that the standard spectral line shape codes as FLYCHK are not adapted to reproduce these situations. The results presented in this review can be used for a better understanding of intense laser–plasma interactions and for laboratory modelling of physical processes in astrophysical objects.

Author Contributions: All authors contributed equally to this review.

Funding: This research received no external fundings.

Acknowledgments: The experimental results were obtained within the state assignment of FASO of Russia to JIHT, RAS (topic #01201357846).

Conflicts of Interest: The authors declare no conflicts of interest.

References

1. Danson, C.N.; Brummitt, P.A.; Clarke, R.J.; Collier, J.L.; Fell, B.; Frackiewicz, A.J.; Hawkes, S.; Hernandez-Gomez, C.; Holligan, P.; Hutchinson, M.H.R.; et al. Vulcan petawatt: Design, operation and interactions at 5.10^{20} Wcm^{-2}. *Laser Part. Beams* **2005**, *23*, 87. [CrossRef]
2. Oks, E. *Plasma Spectroscopy: The Influence of Microwave and Laser Fields*; Ecker, G., Lambropoulos, P., Sobelman, I.I., Walther, H., Lotsch, H.K.V., Eds.; Springer-Verlag: Berlin/Heidelberg, Germany, 1995.
3. Gavrilenko, V.P.; Oks, E. Intra-Stark spectroscopy of Coulomb radiators in a plasma with a quasi-monochromatic electric field. *Sov. Phys. J. Plasma Phys.* **1987**, *13*, 22.
4. Oks, E.; Böddeker, S.; Kunze, H.J. Spectroscopy of atomic hydrogen in dense plasmas in the presence of dynamic fields: Intra-Stark Spectroscopy. *Phys. Rev. A* **1991**, *44*, 8338. [CrossRef] [PubMed]
5. Gavrilenlo, V.P.; Oks, E. New effect in the Stark spectroscopy of atomic hydrogen: Dynamic resonance. *Sov. Phys. JETP* **1981**, *53*, 1122.
6. Dalimier, E.; Oks, E.; Renner, O. Review of langmuir-wave-caused dips and charge-exchange caused dips in spectral lines from plasmas and their applications. *Atoms* **2014**, *2*, 178–194. [CrossRef]
7. Dalimier, E.; Faenov, A.Y.; Oks, E.; Angelo, P.; Pikuz, T.A.; Fukuda, Y.; Andreev, A.; Koga, J.; Sakaki, H.; Kotaki, H.; et al. X-ray spectroscopy of super intense laser-produced plasmas for the study of non-linear processes. Comparison with PIC simulations. *J. Phys. Conf. Ser.* **2017**, *810*, 012004. [CrossRef]
8. Dalimier, E.; Oks, E.; Renner, O. Dips in spectral line profiles and their applications in plasma physics and atomic physics. *AIP Conf. Proc.* **2017**, *1811*, 190003.
9. Lu, J.; Xiao, S.; Yang, Q.; Liu, L.; Wu, Y. Spatially-resolved spectra from a new uniform dispersion crystal spectrometer for characterisation of Z-pinch plasmas. *J. Quant. Spectrosc. Radiat. Transf.* **2013**, *116*, 41.
10. Renner, O.; Dalimier, E.; Oks, E.; Krasniqi, F.; Dufour, E.; Schott, R.; Foerster, E. Experimental evidence of Langmuir-wave caused features in spectral lines of laser-produced plasmas. *J. Quant. Spectrosc. Radiat. Transf.* **2006**, *99*, 439. [CrossRef]
11. Oks, E.; Dalimier, E.; Faenov, A.Y.; Pikuz, T.; Fukuda, Y.; Jinno, S.; Sakaki, H.; Kotaki, H.; Pirozhkov, A.; Hayashi, Y.; et al. Two-plasmon decay instability's signature in spectral lines and spectroscopic measurements of charge exchange rate in femtosecond laser-driven cluster-based plasma. *J. Phys. B At. Mol. Opt. Phys.* **2014**, *47*, 221001. [CrossRef]

12. Oks, E.; Dalimier, E.; Faenov, A.Y.; Angelo, P.; Pikuz, S.A.; Tubman, E.; Butler, N.M.H.; Dance, R.J.; Pikuz, T.A.; Skobelev, I.Y.; et al. Using X-ray spectroscopy of relativistic laser plasma interaction to reveal parametric decay instabilities: A modeling tool for astrophysics. *Opt. Express* **2017**, *25*, 1958. [CrossRef] [PubMed]
13. Oks, E.; Dalimier, E.; Faenov, A.Y.; Angelo, P.; Pikuz, S.A.; Pikuz, T.A.; Skobelev, I.Y.; Ryazanzev, S.N.; Durey, P.; Doehm, L.; et al. In-depth study of intra-Stark spectroscopy in the X-ray range in relativistic laser-plasma interactions. *J. Phys. B At. Mol. Opt. Phys.* **2017**, *50*, 245006. [CrossRef]
14. Depierreux, S.; Fuchs, J.; Labaune, C.; Michard, A.; Baldis, H.A.; Pesme, D.; Hüller, S.; Laval, G. First observation of ion acoustic waves produced by the langmuir decay instability. *Phys. Rev. Lett.* **2000**, *84*, 2869. [CrossRef] [PubMed]
15. Dromey, B.; Kar, S.; Bellei, C.; Carroll, D.C.; Clarke, R.J.; Green, J.S.; Kneip, S.; Markey, K.; Nagel, S.R.; Simpson, P.T.; et al. Bright multi-keV harmonic generation from relativistically oscillating plasma surfaces. *Phys. Rev. Lett.* **2007**, *99*, 085001. [CrossRef] [PubMed]
16. Danson, C.N.; Brummitt, P.A.; Clarke, R.J.; Collier, J.L.; Fell, B.; Frackiewicz, A.J.; Hancock, S.; Hawkes, S.; Hernandez-Gomez, C.; Holligan, P.; et al. Vulcan Petawatt—An ultra-high-intensity interaction facility. *Nucl. Fusion* **2004**, *44*, 5239. [CrossRef]
17. Dover, N.P.; Palmer, C.A.J.; Streeter, M.J.V.; Ahmed, H.; Albertazzi, B.; Borghesi, M.; Carroll, D.C.; Fuchs, J.; Heathcote, R.; Hilz, P.; et al. Buffered high charge spectrally-peaked proton beams in the relativistic-transparency regime. *New J. Phys.* **2016**, *18*, 013038. [CrossRef]
18. Faenov, A.Y.; Pikuz, S.A.; Erko, A.I.; Bryunetkin, B.A.; Dyakin, V.M.; Ivanenkov, G.V.; Mingaleev, A.R.; Pikuz, T.A.; Romanova, V.M.; Shelkovenko, T.A. High-performance X-ray spectroscopic devices for plasma microsources investigations. *Phys. Scr.* **1994**, *50*, 333. [CrossRef]
19. Lavrinenko, Y.S.; Morozov, I.V.; Pikuz, S.A.; Skobelev, I.Y. Reflectivity and imaging capabilities of spherically bent crystals studied by ray-tracing simulations. *J. Phys. Conf. Ser.* **2015**, *653*, 012027. [CrossRef]
20. Alkhimova, M.A.; Pikuz, S.A.; Faenov, A.Y.; Skobelev, I.Y. Determination of spectral reflectivity of spherically bent mica crystals applied for diagnostics of relativistic laser plasmas. *J. Phys. Conf. Ser.* **2016**, *774*, 01215. [CrossRef]
21. Böddeker, S.; Kunze, H.J.; Oks, E. Novel structure in the H alpha line profile of hydrogen in dense helium plasma. *Phys. Rev. Lett.* **1995**, *75*, 4740. [CrossRef] [PubMed]
22. Maine, P.; Strickland, D.; Bado, P.; Pessot, M.; Mourou, G. Generation of ultrahigh peak power pulses by chirped pulse amplification. *IEEE J. Quant. Electron* **1988**, *QE-24*, 398. [CrossRef]
23. Akhiezer, A.I.; Polovin, R.V. Theory of wave motion of an electron plasma. *Sov. Phys. JETP* **1956**, *3*, 696.
24. Lünow, W. On the relativistic non-linear interaction of cold plasma with electro-magnetic waves. *Plasma Phys.* **1968**, *10*, 879. [CrossRef]
25. Guerin, S.; Mora, P.; Adam, J.C.; Heron, A.; Laval, G. Propagation of ultra-intense laser pulses through over-dense plasma layers. *Phys. Plasmas* **1996**, *3*, 2693. [CrossRef]
26. Gray, R.J.; Carroll, D.C.; Yuan, X.H.; Brenner, C.M.; Burza, M.; Coury, M.; Lancaster, K.L.; Lin, X.X.; Li, Y.T.; Neely, D.; et al. Laser pulse propagation and enhanced energy coupling to fast electrons in dense plasma gradients. *New J. Phys.* **2014**, *16*, 113075. [CrossRef]
27. Mourou, G.M.; Tajima, T.; Bulanov, S.V. Optics in the relativistic regime. *Rev. Modern Phys.* **2006**, *78*, 309. [CrossRef]
28. Oks, E. *Diagnostics of Laboratory and Astrophysical Plasmas Using Spectral Line-Shapes of One-, Two-, and Three-Electron Systems*; Work Scientific Publishing: Singapore, 2017; ISBN 9814699071.
29. Oks, E.; Sholin, G.V. On Stark profiles of hydrogen lines in a plasma with low-frequency turbulence. *Sov. Phys. Tech. Phys.* **1976**, *21*, 144.
30. Iglesias, C.A.; Dewitt, H.E.; Lebowitz, J.L.; MacGowan, D.; Hubbard, W.B. Low-frequency electric microfield distributions in plasmas. *Phys. Rev. A* **1985**, *31*, 1698. [CrossRef]
31. Dalimier, E.; Oks, E. Robust computational method for fast calculations of multi-charged ions line-shapes affected by a low-frequency electrostatic plasma turbulence. *J. Phys. B At. Mol. Opt. Phys.* **2017**, *50*, 025701. [CrossRef]
32. Djurovic, S.; Ćirišan, M.; Demura, A.V.; Demchenko, G.V.; Nikolić, D.; Gigosos, M.A.; González, M.A. Measurements of H_β Stark central asymmetry and its analysis through standard theory and computer simulations. *Phys. Rev. E* **2009**, *79*, 046402. [CrossRef] [PubMed]

33. Demura, A.V.; Demchenko, G.V.; Nikolic, D. Multiparametric dependence of hydrogen Stark profiles asymmetry. *Eur. Phys. J. D* **2008**, *46*, 111. [CrossRef]

34. Chung, H.K.; Chen, M.H.; Morgan, W.L.; Ralchenko, Y.; Lee, R.W. Generalised population kinetics and spectral model for rapid spectroscopic analysis for all elements. *High Energy Density Phys.* **2005**, *1*, 3. [CrossRef]

35. Lichters, R.; Meyer-ter-Vehn, J.; Pukhov, A. Short pulse laser harmonics from oscillating plasma surfaces driven at relativistic intensity. *Phys. Plasmas* **1996**, *3*, 3425–3437. [CrossRef]

36. Kruer, W.L. *The Physics of Laser Plasma Interactions*; Edition Reprint, Collection Frontiers in Physics; Taylor and Francis, Ed.; Westview Press: Boulder, CO, USA, 2003.

atoms

MDPI

Article

Interaction of Ultrashort Laser Pulses with Atoms in Plasmas

V. A. Astapenko [1,*] and V. S. Lisitsa [1,2,3]

[1]   Moscow Institute of Physics and Technology (State University), 141700 Dolgoprudnyi, Russia;
      vlisitsa@yandex.ru
[2]   National Research Centre, Kurchatov Institute, 123182 Moscow, Russia
[3]   National Research Nuclear University MEPhI, 115409 Moscow, Russia
[*]   Correspondence: astval@mail.ru; Tel.: +7-962-9938904

Received: 7 June 2018; Accepted: 10 July 2018; Published: 11 July 2018

Abstract: The paper is devoted to the investigation of the absorption of ultrashort laser pulses on atoms in plasmas, accounting for the different broadening mechanisms of atomic resonant transitions. The analysis is made in terms of the absorption probability during the entire interaction between the laser pulse and atom. Attention is mainly given to dependence of probability upon the pulse duration and the carrier frequency of the pulse. The results are presented via dimensionless parameters and functions describing the effect of finite pulse duration on atomic spectra for different broadening mechanisms, namely Doppler, Voigt, Holtsmark and their combinations, as well as the Stark line broadening of Rydberg atomic lines.

Keywords: ultra-short laser pulse; photo-absorption; spectral line shape; Stark broadening

1. Introduction

The progress in the generation of ultrashort electromagnetic pulses (USPs) [1–3] means that it is important to consider their interaction with atomic systems in plasmas. One of the essential advantages of USPs is connected with their short wavelengths, which make it possible for them to excite atomic systems from ground states, in contrast with femtosecond laser pulses in visible spectral ranges.

The interaction of USPs with atomic systems is of great interest for plasma investigations. The long laser pulses in standard diagnostics (fluorescence schemes) usually deal with the excitation of radiative transitions between excited atomic states which belong to the visible spectral range. The contrast of the fluorescence signal in this case is rather small because the populations of excited atomic levels are not very different from one another under standard plasma conditions. At the same time, the carrier frequency of a USP is often in the ultraviolet or even X-ray spectral range, enabling the excitation of atoms and ions in plasmas from their ground states. In this case, the observed fluorescence signal grows by many orders of magnitude due to the large populations of ground states. This opens up new possibilities for plasma diagnostics (see the discussions in [4]).

In this sense, the X-ray radiation of free electron lasers is of special interest, providing possibilities for both highly charged ion excitation from their ground states and for the tuning of laser frequencies [1]. Moreover, the wide frequency spectra of USPs provide the possibility of USP radiation penetrating into optically dense media.

A specific USP results in the appearance of a new parameter in line broadening theory, namely the pulse duration; see also [5–7]. This must be considered together with standard line shape parameters such as frequency detuning from atomic transition resonance [8,9]. Thus, one deals with a generalization of the standard broadening theory by introducing new line shapes which depend on both frequency and pulse duration. It is clear that for long pulses, the new line shapes make a smooth transition to the standard one, whereas for short pulses, absorption probabilities as nonlinear functions of pulse duration can be observed.

The USP can be used for plasma investigations both in X-ray and visible spectral ranges. The pulses provided by X-ray free electron lasers (XFEL) are useful for investigations of high temperature, dense plasmas where the radiative transitions in atomic systems such as highly charged ions belong to the X-ray spectral range. At the same time, a USP in the visible spectral range (HHG pulses) can be applied for investigations of low temperature plasmas, where radiative transitions in neutral atoms (for example, alkali atoms, etc.) are only in the visible spectral range. The expressions for radiative transition probabilities in this paper are expressed in terms of universal functions, enabling their application for both of the cases mentioned above. Moreover, such universal representation opens up the possibility of transferring the results of specific investigations to other conditions following the dimensionless parameters of the research.

The present paper is dedicated to the analysis of the absorption of USPs on atoms in plasmas, accounting for different broadening mechanisms, including the Stark broadening of the atomic resonant transition in terms of total absorption probability during the entire pulse action. Specific attention is devoted to the investigations of absorption probability transitions from the nonlinear to the standard linear absorption regime. This research is presented in terms of universal dimensionless parameters, making it possible to estimate the pulse duration effect on the absorption probabilities for different broadening mechanisms related to different plasma conditions.

This research is made within the frame of first-order perturbation theory with respect to pulse amplitude E_0. Thus, the absorption probability is proportional to the squared product of E_0 and pulse duration τ. The probability is therefore quite different for XFEL and HHG pulses. A specific example is considered in [10].

The structure of the paper is as follows: Section 2 contains general formulas for USP absorption line shapes, presented in terms of universal broadening parameters; Section 3 is devoted to the application of general study of the specific broadening mechanisms, namely Doppler, Voigt, Holtsmark and Rydberg spectra; and Section 4 presents the conclusions.

2. General Formulas

The total absorption probability for laser pulses with a Gaussian envelope in dimensionless variables takes the form [7]

$$W(\alpha,\delta) = \frac{\pi}{4}\frac{f_0\,E_0^2}{\omega_0}\frac{\alpha^2}{\Delta}\int_{-\infty}^{\infty}\exp\left\{-\alpha^2\,(\beta-\delta)^2\right\}G(\beta)d\beta = \frac{\pi^{3/2}}{4}\frac{f_0\,E_0^2}{\omega_0}\frac{1}{\Delta}F(\alpha,\delta) \tag{1}$$

where $G(\omega')$ is the spectral line shape of a radiative transition between atomic states. Here, dimensionless variables are introduced:

$$\beta = \frac{\omega'-\omega_0}{\Delta},\delta = \frac{\omega-\omega_0}{\Delta},\alpha = \Delta\tau \tag{2}$$

where Δ is the spectral width of the line, ω_0 and f_0 are the frequency and oscillator strength of the electron transition, E_0 is the amplitude of electric field strength of the USP, and ω and τ are the carrier frequency and pulse duration. The dimensionless absorption probability $F(\alpha,\delta)$ is given by the following expression:

$$F(\alpha,\delta) = \frac{\alpha^2}{\sqrt{\pi}}\int_{-\infty}^{\infty}\exp\left\{-\alpha^2\,(\beta-\delta)^2\right\}G(\beta)\,d\beta \tag{3}$$

This function describes the dependence of the absorption probability on pulse duration and carrier frequency in terms of dimensionless parameters α and δ. Note that

$$F(\alpha,\delta)/\alpha \to G(\delta) \text{ for } \alpha \to \infty \tag{4}$$

Thus, the function $F(\alpha, \delta)$ divided by parameter α presents the line shape of the USP absorption modified by finite pulse duration and becoming the standard one when the value of α goes to infinity.

It should be mentioned that the function $F(\alpha, \delta)/\alpha$ integrated over detuning parameter δ is equal to unity.

The influence of pulse duration parameter $\alpha = \Delta\tau$ on absorption probability for different mechanisms of line broadening in plasmas is considered below.

3. Results and Discussions

3.1. Doppler Broadening

In the case of Doppler broadening, the integral in expression (1) can be calculated analytically, and Formula (1) turns into

$$W_D = \frac{\pi^{3/2}}{4} \frac{f_0 E_0^2}{\omega_0} \frac{1}{\Delta\omega_D^2} F_D(\alpha, \delta) \tag{5}$$

where

$$F_D(\alpha, \delta) = \frac{1}{\sqrt{\pi}} \frac{\alpha^2}{\sqrt{2\alpha^2 + 1}} \exp\left\{-\frac{\alpha^2 \delta^2}{2\alpha^2 + 1}\right\} \tag{6}$$

is the function describing the excitation probability depending on the dimensionless parameters α, and δ. The line shape width Δ in this case is equal to the standard Doppler line width; see [8]. One can conclude from Equation (6) that

$$\int F_D(\alpha, \delta)/\alpha \, d\delta = 1 \tag{7}$$

for any α. Thus, the normalized line shape has the form

$$\tilde{G}_D(\alpha, \delta) = F_D(\alpha, \delta)/\alpha = \frac{1}{\sqrt{\pi}\,\sigma(\alpha)} \exp\left(-\frac{\delta^2}{2\sigma^2(\alpha)}\right) \tag{8}$$

where

$$\sigma(\alpha) = 2^{-1/2} \frac{\sqrt{2\alpha^2 + 1}}{\alpha} \tag{9}$$

is the modified spectral width for Doppler broadening and Gaussian laser pulse. Note that for a large pulse duration, we have $\sigma(\alpha \gg 1) \to 1$, which is the standard Doppler line shape.

The Doppler function $F_D(\alpha, \delta)$ has extrema which are dependent on the pulse duration parameter $\alpha = \Delta\tau$ starting from some critical value of the detuning parameter $\delta = (\omega - \omega_0)/\Delta$. Critical dimensionless detuning, δ^*, is equal to

$$\delta^* = \sqrt{3 + 2\sqrt{2}} \cong 2.414 \tag{10}$$

For $|\delta| < \delta^*$, extrema in the $F_D(\alpha)$ function are absent. In the opposite case, the absorption probability dependent on pulse duration has a maximum and minimum, as one can see from Figure 1. For sufficiently long pulses ($\alpha > 1$), the dependence of $F_D(\alpha)$ becomes linear, as is the case in the standard approach for long pulse durations.

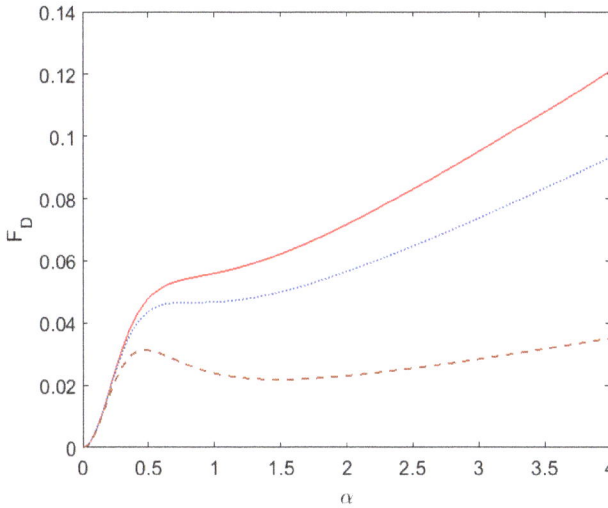

Figure 1. Dependence of Doppler function F_D upon dimensionless pulse duration for different values of dimensionless detuning δ: solid line $- \delta = 2.3$, dotted line $- \delta = 2.414$, dashed line $- \delta = 2.8$.

Figure 2 presents the distortion of the Doppler line shape by finite duration times of USP. Here, the absorption line shape $F_D(\alpha, \delta)/\alpha$ is shown for different values of α. A smooth increase of the Gaussian line shape (8) can be seen in accordance with the decrease of the dispersion factor (9).

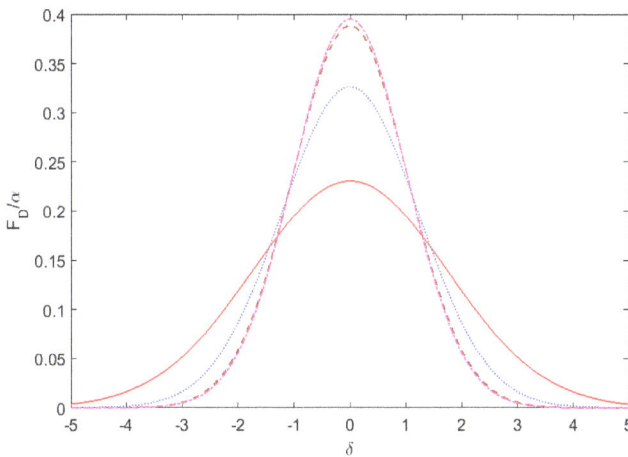

Figure 2. Absorption line shape $F_D(\alpha, \delta)/\alpha$ for different values of dimensionless pulse duration; α: solid line $- \alpha = 0.5$, dotted line $- \alpha = 1$, dashed line $- \alpha = 3$, dotted–dashed line $- \alpha = 5$.

3.2. Voigt Line Shape

The Doppler line shape presents the composition of two Gaussian spectral distributions, namely the Doppler line shape and Gaussian shape of USP. Let us consider the important case of the Voigt line shape, which is a composition of two mechanisms of line broadening, namely the Doppler and collisional (impact) mechanisms [8,9]. In this case, we should substitute the Voigt line shape $G_V(\beta, \rho = \gamma/\Delta)$ on the right side of (1) and (3). Therefore, the probability of absorption is a function of three arguments: $F_V(\alpha, \delta, \rho)$ where $\rho = \gamma/\Delta$ is the ratio between Lorentzian and Gaussian spectral widths, being the standard parameter of Voigt line shapes.

The effect of pulse duration parameter, α, on the transition probability for different ratios, ρ, is presented in Figure 3. It is shown that the transition to the linear dependence takes place for a lower value of the parameter ρ with the increase of the contribution of Lorentzian line wings in the Voigt spectral shape.

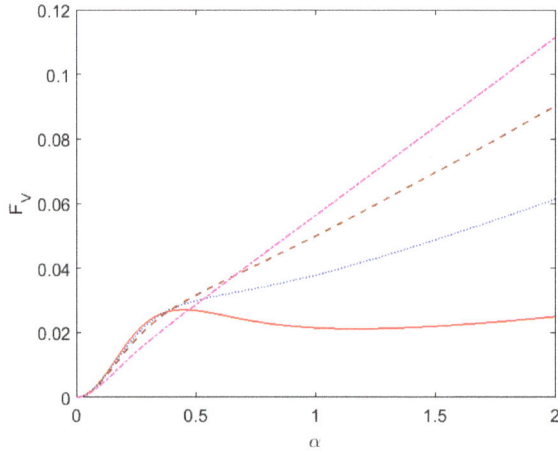

Figure 3. Dependence of Voigt function F_V upon dimensionless pulse duration for dimensionless detuning $\delta = 3$ and different values of ratio $\rho = \gamma/\Delta$: solid line – $\rho = 0.1$, dotted line – $\rho = 0.5$, dashed line – $\rho = 1$, dotted–dashed line – $\rho = 3$.

3.3. Holtsmark Line Shape

Holtsmark line shapes are the basis for the description of hydrogen spectral line broadening in plasmas [8]. The specifics of such static line shapes are the differences between lines with and without central Stark components, having corresponding peaks and dips in the center of the spectral line. The shape of the unshifted central Stark component is well described by a Voigt line shape with electron impact width. We demonstrate the effect of finite pulse duration on the static line shapes without central components described by Holtsmark distribution.

The Holtsmark distribution has to be substituted into Equations (1) and (3) accounting for a standard static line width $\Delta = CF_0$, where C is the effective Stark constant for the line under consideration and F_0 (a.u.) $= 2.6\ N^{2/3}$ (N is ion density) is the standard ion field strength in the Holtsmark distribution. The effect of finite pulse duration α on the Holtsmark static line shape is presented in Figure 4 for different values of detuning δ. This figure shows that, in the case of zero detuning, the line shape determining the absorption probability goes to zero—no absorption in the line center because of the zero in the line shape center. Thus, in the case of nonzero detunings, the absorption probability goes to the standard line dependence on pulse duration for sufficiently long pulses ($\alpha > 1$).

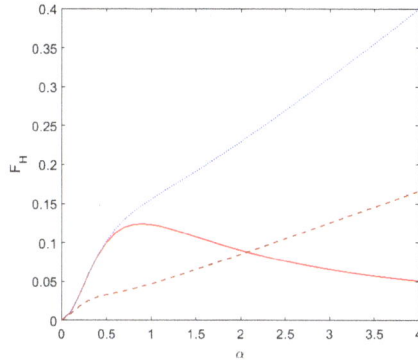

Figure 4. Dependence of Holtsmark function F_H upon pulse duration parameter α for different values of dimensionless detuning δ: solid line $- \delta = 0$, dotted line $- \delta = 0.5$, dashed line $- \delta = 5$.

Figure 5 demonstrates the effect of pulse duration on the Holtsmark–Doppler generalized line shape $F_{HD}(\alpha, \delta, \zeta = \Delta_D / \Delta_H)$ for different values of Doppler to Holtsmark line width ratios, ζ, and pulse durations, α. It is seen that the effect of the Doppler broadening produces a decrease of the dip in the center of the static Holtsmark line shape.

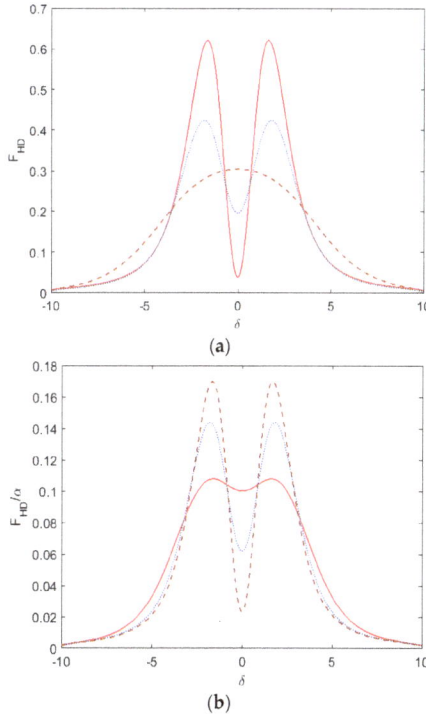

(a)

(b)

Figure 5. Holtsmark–Doppler absorption line shape F_{HD} for different values of Doppler to Stark widths ζ: (a) solid line $- \zeta = 0.05$, dotted line $- \zeta = 1$, dashed line $- \zeta = 3$; $\alpha = 3$ and for different pulse duration parameters; α: (b) solid line $- \alpha = 0.5$, dotted line $- \alpha = 1$, dashed line $- \alpha = 2$; $\zeta = 0.1$.

It is seen from the comparison of Figure 5a,b that the pulse duration produces approximately the same effect on the static line shapes as Doppler broadening.

The dependences of Holtsmark–Doppler absorption line shapes on the pulse duration parameter, α, for zero detuning and the different ratios of Doppler to Holtsmark widths, ζ, are shown in Figure 6.

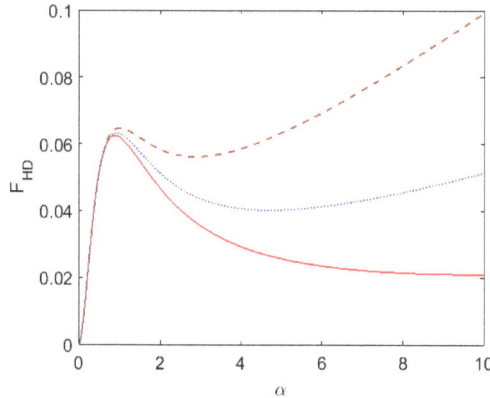

Figure 6. Stark–Doppler absorption probability F_{HD} as a function of pulse duration parameter α for zero detuning $\delta = 0$ and various values of parameter ζ: solid line – $\zeta = 0.1$, dotted line – $\zeta = 0.2$, dashed line – $\zeta = 0.3$.

A smooth transition of the absorption probability to the linear dependence on pulse duration and a reduction of the absorption probability magnitude at small values of the parameter ζ due to the small value of the Holtsmark line shape in the central part of the spectral line can be observed.

3.4. Line Shapes of Highly Excited Rydberg Atomic States.

Hydrogen line shapes for radiative transitions from highly exited (Rydberg) states to low-excited ones hasvea simplified static line shape, due to the slow variation of the sum of π and σ component intensities for such transitions [11]. This makes it possible to present the corresponding line shapes in a universal manner, taking into account simultaneously both static ion and impact electron contributions as well as Stark–Doppler broadening. We use these results to calculate the USP absorption probability $F_R(\alpha, \delta)$ by the substitution of corresponding shapes from [11] into the general Equation (3).

Such a static Stark line shape in this approximation accounting for electron collision broadening γ has the form [11]

$$G_{RL}(\beta, \gamma) = \frac{1}{\pi} \int_0^\infty \cos(\beta x) \exp\left(-\gamma x - x^{3/2}\right) \tag{11}$$

where β is the standard line shape parameter and γ is the ratio of the impact line width to the static width. The effect of pulse duration on absorption probability $F_{RL}(\alpha, \delta, \gamma)$ is shown in Figure 7 for detuning parameter $\delta = 5$ and different relative values of impact to static line widths. It is clear that the transition from static to impact line shape following the increase of the impact width γ facilitates the transition from a linear to nonlinear regime in USP absorption.

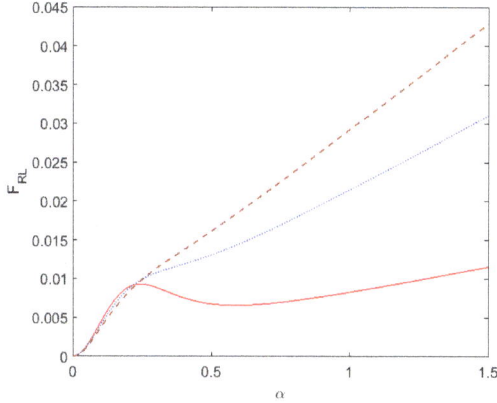

Figure 7. Stark–Lorentz absorption probability F_{RL} for Rydberg radiative transitions in H-atoms as a function of pulse duration parameter α, detuning parameter $\delta = 5$ and different ratios of impact collision to static widths: solid line $- \gamma = 0$, dotted line $- \gamma = 1$, dashed line $- \gamma = 2$.

The same procedure can be done with account for Stark–Doppler broadening:

$$G_{RD}(\beta, \sigma) = \frac{1}{\pi} \int_0^{\infty} \cos(\beta x) \exp\left(-\sigma^2 x^2/2 - x^{3/2}\right) \qquad (12)$$

The effect of pulse duration on the Rydberg Stark–Doppler absorption probability $F_{RD}(\alpha, \delta, \sigma)$ is shown in Figure 8. It is clear that the minimum in such dependence disappears with the increase of Doppler broadening. A comparison of Figures 7 and 8 shows that the Lorentz broadening affects the dependence of the function $F(\alpha)$ more strongly than the Doppler one due to the wider wings in the spectral dependence of the Lorentzian broadening.

From the comparison of Figures 7 and 8 with previous figures describing Doppler, Voigt and Holtsmark line shapes, one can see that, for Rydberg absorption, the function $F(\alpha)$ reaches its maximum at a smaller value of dimensionless pulse duration ($\alpha_{max} \approx 0.25$), while in other cases $\alpha_{max} \approx 0.5$.

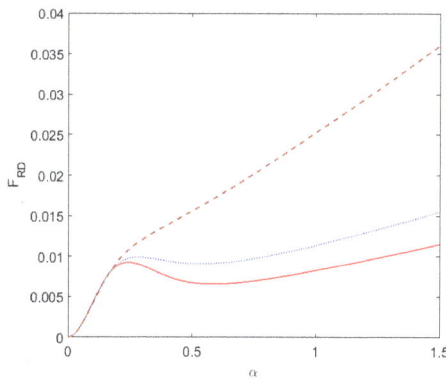

Figure 8. The Stark–Doppler absorption probability F_{RD} for Rydberg radiative transitions in H-atoms as a function of pulse duration parameter α, detuning parameter $\delta = 5$ and different ratio σ of Doppler to static widths: solid line $- \sigma = 0$, dotted line $- \sigma = 1$, dashed line $- \sigma = 2$.

The line shapes of radiative transitions are strongly simplified for Rydberg atomic states, making it possible to take into account Stark, collisional and Doppler broadening simultaneously according to equations (11–12). This also simplifies the research of the effect of USPs on atomic transitions for Rydberg lines.

4. Conclusions

The calculations of absorption probabilities in the interaction of USPs with atomic systems in plasmas are made within the frame of perturbation theory regarding the amplitude of laser pulses. The results are expressed in terms of generalized line shapes of the absorption, which contain additional parameters related to pulse duration together with other standard parameters of the broadening theory. Specific broadening mechanisms relative to plasma conditions are considered: Doppler, Voigt, Holtsmark, and their combinations, as well as Stark line broadening of Rydberg atomic lines. The effects of pulse duration on absorption probabilities as well as their spectral dependence are investigated in detail. The specific interest of the research is the transition of the absorption probabilities from nonlinear time dependence at a small pulse duration to linear ones for long pulses depending on carrier frequency detuning from their frequency of transition. The research is presented in universal form, making applications for different atoms and plasma conditions possible.

Author Contributions: All work has been done by the authors themselves. V.S. suggested the problem and wrote the paper, V.A. performed the calculations and drew figures.

Funding: This research received no external funding.

Acknowledgments: The research was supported by the government order of the Ministry of Education and Science of the Russian Federation (Project No. 3.9890.2017/8.9).

Conflicts of Interest: The authors declare no conflict of interest.

References

1. Tanaka, T. Proposal to generate an isolated monocycle X-ray pulse by counteracting the slippage effect in free-electron lasers. *Phys. Rev. Lett.* **2015**, *114*, 044801. [CrossRef] [PubMed]
2. Kida, Y.; Kinjo, R.; Tanaka, T. Synthesizing high-order harmonics to generate a sub-cycle pulse in free-electron lasers. *Appl. Phys. Lett.* **2016**, *109*, 151107. [CrossRef]
3. Chini, M.; Zhao, K.; Chang, Z. The generation, characterization and applications of broadband isolated attosecond pulses. *Nat. Photonics* **2014**, *8*, 178–186. [CrossRef]
4. Back, C.; Lee, R.; Chenais-Popovics, C. Measurement of resonance fluorescence in a laser produced Al XII plasma. *Phys. Rev. Lett.* **1989**, *63*, 1471–1474. [CrossRef] [PubMed]
5. Astapenko, V.A. Simple formula for photoprocesses in ultrashort electromagnetic field. *Phys. Lett. A* **2010**, *374*, 1585–1590. [CrossRef]
6. Astapenko, V.A. Scattering of ultrashort electromagnetic radiation pulse by an atom in a broad spectral range. *J. Exp. Theor. Phys.* **2011**, *112*, 193–198. [CrossRef]
7. Calisti, A.; Astapenko, V.A.; Lisitsa, V.S. Excitation of hydrogen atom by ultrashort laser pulses in optically dense plasma. *Contrib. Plasma Phys.* **2017**, *57*, 414–420. [CrossRef]
8. Griem, H.R. *Spectral Line Broadening by Plasmas*; Academic Press: New York, NY, USA, 1974.
9. Sobelman, I.I. *Introduction to the Theory of Atomic Spectra*; Pergamon Press: Oxford, UK; New York, NY, USA, 1972.
10. Astapenko, V.A.; Bagan, V.A. Features of excitation of a two-level system by short nonresonance laser pulses. *J. Phys. Sci. Appl.* **2013**, *3*, 269–277.
11. Stambulchik, E.; Maron, Y. Stark effect of high-n hydrogen-like transitions: Quasi-contiguous approximation. *J. Phys. B At. Mol. Opt. Phys.* **2008**, *41*, 095703. [CrossRef]

atoms

|MDPI|

Article

Measurement of Electron Density from Stark-Broadened Spectral Lines Appearing in Silver Nanomaterial Plasma

Ashraf M. EL Sherbini [1], Ahmed E. EL Sherbini [1] and Christian G. Parigger [2,*]

[1] Laboratory of Laser and New Materials, Faculty of Science, Cairo University, Giza 12613, Egypt;
     elsherbiniaa@gmail.com (A.M.EL.S.); ahmed.azdakyne@gmail.com (A.E.EL.S.)
[2] Department of Physics and Astronomy, University of Tennessee/University of Tennessee Space Institute,
     411 B. H. Goethert Parkway, Tullahoma, TN 37388, USA
*   Correspondence: cparigge@tennessee.edu; Tel.: +1-931-841-5690

Received: 31 July 2018; Accepted: 8 August 2018; Published: 13 August 2018

Abstract: This work communicates results from optical emission spectroscopy following laser-induced optical breakdown at or near nanomaterial. Selected atomic lines of silver are evaluated for a consistent determination of electron density. Comparisons are presented with Balmer series hydrogen results. Measurements free of self-absorption effects are of particular interest. For several silver lines, asymmetries are observed in the recorded line profiles. Electron densities of interest range from 0.5 to 3×10^{17} cm^{-3} for five nanosecond Q-switched Nd:YAG radiation at wavelengths of 1064 nm, 532 nm, and 355 nm and for selected silver emission lines including 328.06 nm, 338.28 nm, 768.7 nm, and 827.3 nm and the hydrogen alpha Balmer series line at 656.3 nm. Line asymmetries are presented for the 328.06-nm and 338.28-nm Ag I lines that are measured following generation of the plasma due to multiple photon absorption. This work explores electron density variations for different irradiance levels and reports spectral line asymmetry of resonance lines for different laser fluence levels.

Keywords: laser-induced plasma; Stark broadening; electron density; nanomaterial; atomic spectroscopy; silver; hydrogen

---

1. Introduction

Laser-induced breakdown spectroscopy (LIBS) [1] is utilized for measuring plasma that is generated at or near silver nanomaterial. During the last few decades, LIBS has been recognized for its versatility and integral aspect for a variety of spectro-chemical analysis procedures. Typically, high peak power and nominal nanosecond radiation is focused to irradiance levels of the order of a few MW/cm^2 to TW/cm^2 and the emitted light is recorded with spectrometers and gated array detectors [2,3]. Historically, laser-induced plasma spectroscopy (LIPS) explores the physics of the plasma induced by laser light via optical emission spectroscopy (OES) [4–7]. Spectral line shape analysis via OES leads to the determination of at least one characteristic plasma parameter such as electron density, n_e.

The measurement of electron density is of prime importance for the description of the plasma induced by laser radiation. Spectroscopically, n_e can be measured using different experimental techniques that include the measurement of the optical refractivity of the plasma [4–7], calculation of the principal quantum number in the series limit [4–7], measurement of the absolute emission coefficient (spectral radiance per wavelength in W sr^{-1} m^{-3}) of a spectral line [8], and measurement of the absolute emissivity of the continuum emission (W sr^{-1} m^{-3}) [8]. However, measurement of Stark broadening of emitted lines for n_e determination has been widely utilized [4–8].

Measurements of n_e from Stark broadening is relatively straightforward provided that the Stark effect is the dominant broadening mechanism with significantly smaller contributions from Doppler broadening and other pressure broadening mechanisms resulting from collisions with neutral atoms (i.e., resonance and Van der Waals broadening) [4–7]. Theoretical calculations of Stark broadening parameters of hydrogen and hydrogenic lines are communicated by H. Griem [4] and E. Oks [9–11]. Precise fitting of the measured line shapes to convolutions of Lorentzian and Gaussian spectral line shapes (i.e., Voigt line shapes) allows one to extract the Stark full width at half maximum (FWHM). Subsequently, the electron density can be inferred from tabulated Stark broadening tables.

2. Nanomaterial

Nanomaterials usually describe structured components with at least one dimension less than 100 nm [12]. Two principal factors cause the properties of nanomaterials to differ significantly from bulk materials such as the increase in the relative surface area and the quantum effects. These factors can substantially change and/or enhance the well-known bulk properties such as chemical reactivity [13], mechanical strength [14], electrical and magnetic [15], and optical characteristics [16]. As the particle size decreases, a greater proportion of atoms are found at the surface than in the interior [17]. The quantum effects can begin to dominate the properties of matter as its size is reduced to the nano-scale. Nanoparticles are of interest because of their inherent new properties when compared with larger particles of the same materials [12–17].

It was found that the addition of a thin layer of gold and silver nanoparticles to the surface of an analyte matrix alloys can lead to signal improvement and, therefore, an improved limit of detection (LOD) in LIBS applications [18]. The acronym associated with improved emission signals is Nano-Enhanced Laser Induced Breakdown Spectroscopy (NELIBS). Conversely, interaction of high peak power radiation with pure nanomaterial targets [19–22] is investigated with so-called Nano-Enhanced Laser Induced Plasma spectroscopy (NELIPS).

In previous NELIPS work, the signal enhancement shows the following trends: (1) the enhanced emission from the nanomaterials increases linearly with time delays when compared with bulk material [19], (2) the enhanced emission increases with decreasing laser fluence [20], (3) there are no apparent changes of the plasma electron density and temperature [21,22], (4) the enhancement factors that may vary for different experimental conditions can be associated with the relative masses ejected from the targets [21,22], (5) the threshold of the plasma ignition from the surface of the nanomaterials is much smaller than that from the corresponding bulk [21,22], (6) the breakdown threshold is inversely proportional to the square of the incident laser wavelength [20–22], and, lastly, (7) the threshold of the plasma from the nano-material targets changes linearly with the diameter size of the nanoparticles [22].

Moreover, the modeling of the laser induced plasma from either type of targets (Bulk and Nano) has been theoretically investigated after the addition of a laser wavelength dependent term [21,22], which was found to contribute by 90% when using near UV laser wavelengths [21,22].

3. Materials and Methods

This work utilizes the same experimental setup reported in previous studies [21,22]. It is comprised of an Nd-YAG laser device (type Quantel-Brilliant B, France [19–22]) operated at the fundamental wavelength of 1064 nm and two higher harmonics at 532 nm and 355 nm with output laser energy of 30 ± 3 mJ, 100 ± 4 mJ, and 370 ± 5 mJ, respectively. The focusing lens was located at a distance about 95 ± 1 mm, which is away from the target material. Using a special thermal paper (supplied by Quantel®, France [19–22]), a circular laser beam spot revealed a radius of 0.27 ± 0.03 mm. In order to avoid laser focusing-lens chromatic aberrations, the plasma initiation was first observed in laboratory air and, subsequently, the target was displaced closer to the 100 mm focal-length achromatic lens. This routine would indicate that the plasma emission originated from the target rather than from ambient air surrounding the target. The light from the plasmas was collected using a 400 μm diameter optical fiber (with numerical aperture NA = 0.22) to the entrance slit of the SE200-Echelle type spectrograph (Catalina Scientific, Tucson, AZ, USA [19–22]) with optical resolution of 0.02 nm per pixel with an average instrumental bandwidth of 0.2 nm. The optical fiber was positioned at a distance of 5 mm from the laser-plasma axis with a precise xyz-holder. The resolved spectra were monitored using a fast response intensified charge-coupled device (ICCD) (type Andor-iStar DH734-18F, Belfast, Northern Ireland [19–22]) and the data acquisition was carried out using KestrelSpec® 3.96 software Catalina Scientific, Tucson, AZ, USA [19–22]) at a resolution of 0.02 nm per pixel (of size 196 µm²).

The nano silver was supplied as a powder (MKNano®, Toronto, ON, CA [19–22]) with the product label MKN-Ag-090, CAS Number 7440-22-4, and with an average size of 90 ± 10 nm. The nano-powder is compressed to circular disk tablets with a diameter of 10 mm using a 500 kg/cm² mechanical press. The shape of the nanoparticles was investigated with transmission electron microscopy (TEM) after compression. Nearly spherical diameters of 95 ± 15 nm are found and only slight distortions were observed. Both delay and gate times were adjusted to the levels of 2 µs across the experimental studies. Background stray light during experimental runs was measured and subtracted with the help of Andor iStar ICCD-KestrelSpec® software (Catalina Scientific, Tucson, AZ, USA [19–22]). The noise level from the detection electronics was recorded across the entire wavelength region (250–850 nm) and was found to be about 20 ± 7 counts. The signal-to-noise ratio was computed using noise-level as the sum of the electronic noise in addition to the continuum emission (sometimes called background radiation) that occurs underneath the atomic lines of interest. The incident laser energy for each laser pulse was measured utilizing a quartz beam splitter. The reflected part (4%) was incident on an absolutely calibrated power-meter (Ophier model 1z02165, North Logan, UT, USA [19–22]). The laser pulse shape was measured using a 25 ps, fast-response photodiode in conjunction with a digital storage oscilloscope (Tektronix model TDS-1012, Tektronix, Beaverton, OR, USA [19–22]) and the pulse-width was found stable at a level of 5 ± 1 ns. The laser energy was adjusted with a set of calibrated neutral density filters. The absolute sensitivity of the spectrograph, camera, and optical fiber was calibrated using a DH2000-CAL lamp (supplied by Ocean Optics-SN: 037990037, Ostfildern, DE [19–22]). The data presented in this paper are taken as the average over three consecutive shots onto fresh targets and the data are presented together with standard deviations about means and plotted as error bars associated with the measurement points. The observed spectral Ag I lines (e.g., 327.9 nm, 338.2 nm, 405.5 nm, 421.2 nm, 447.6 nm, 467.7 nm, 520.9 nm, 546.5 nm, 768.7 nm, and 827.3 nm) were examined in view of self-absorption and/or self-reversal.

4. Results

The spectral emission is recorded in different spectral regions from the plasma that is generated by the interaction of high peak power laser radiation using the third harmonic, blue wavelength of 355 nm with silver nano-based targets and the corresponding bulk material. Figure 1 displays the measured data.

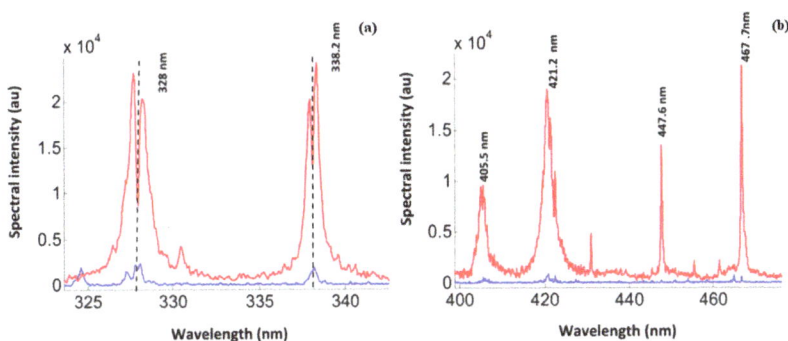

Figure 1. *Cont.*

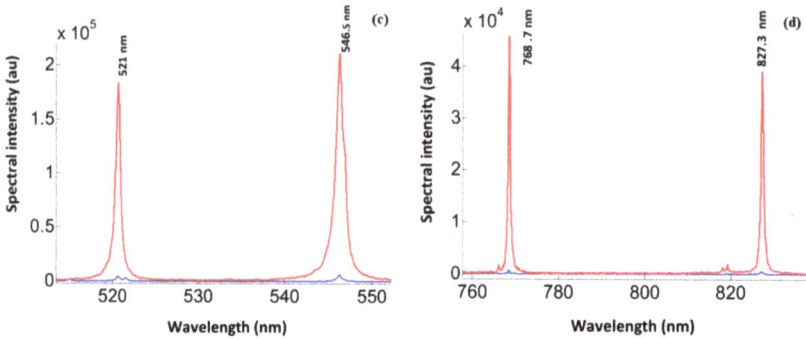

Figure 1. (**a–d**) Emission spectra at different wavelength regions from nano-based silver target (in red) and bulk target plasma (in blue).

There is a clear larger emission from the nano-based silver plasma than from plasma created from the bulk target (ratio of the red to blue curves). Detailed inspection of the resonant transitions ($4d^{10}5p$-$4d^{10}5s$) at wavelengths of 327.9 nm and 338.2 nm (as depicted in Figure 2) indicates the existence of self-reversal as well as self-absorption.

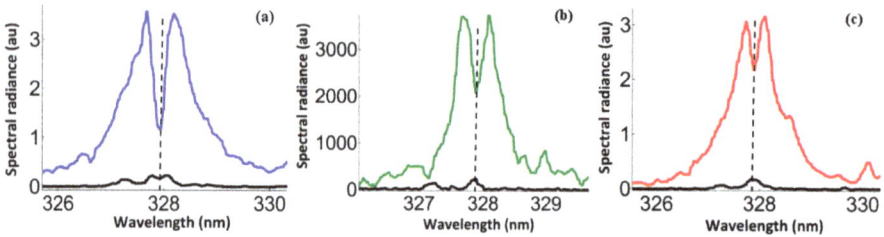

Figure 2. Self-reversal of the resonance Ag I line at 327.9 nm. Laser wavelengths: (**a**) 355 nm, (**b**) 532 nm, and (**c**) 1064 nm. The colored, larger, black, and lower plasma spectra are measured from nano-based and bulk silver targets, respectively.

Self-reversal is often associated with the population density of the ground state of the silver atoms ($4d^{10}5s$-state). The ground-state population exhibits a strong gradient of the plasma parameters (electron density and temperature) ranging from the plasma core to the periphery [1–5]. Moreover, this effect was found to be pronounced at a shorter wavelength laser irradiation of 355 nm (see Figure 2a—blue curve). This is in contrast with the emission from the bulk-based silver target under similar conditions.

The results are consistent with previous studies [21,22] that discussed that the population density of the ground state is larger for the plasma created at the surface of the nano-based target than for the bulk-based plasma, which is described by the enhancement factor defined by $\left(Enh.Factor = \frac{I_0^{Nano}}{I_0^{Bulk}} \approx \frac{N_0^{Nano}}{N_0^{Bulk}} \right)$. In this scenario, $\frac{I_0^{Nano}}{I_0^{Bulk}}$ is the ratio of the spectral radiance of nano-based to bulk-based target plasma spectral line. However, $\frac{I_0^{Nano}}{I_0^{Bulk}}$ is the ratio between the corresponding population density of the ground state of the silver atoms. The quantification of self-absorption and/or self-reversal that one should rely on certain optically thin (standard) spectral lines should allow precise measurements of plasma electron density [23]. The presence of H_α emission spectra provides a good candidate for the measurement of the plasma electron density, but H_α is often absent during the interaction of both green and blue lasers. Therefore, one should consider other optically thin lines that can be measured during the interaction of the different Nd:YAG harmonics with nano-based targets.

The Ag I lines at wavelengths of 768.7 nm and 827.35 nm are candidates for electron density measurements. At the reference density of 1×10^{17} cm^{-3}, the Stark broadening parameters of $w_s^{827.25} = 0.18$ nm and $w_s^{768.7} = 0.17$ nm [24] allow one to estimate electron density from the Lorentzian FWHM component of each line. The fitting of the line shapes to Voigt profiles is shown in Figure 3. The overall results are summarized in Table 1 including electron densities measured from the optically thin H$_\alpha$. In Table 1, the accuracies of n_e are 20%. In this work, the n_e measurement from H$_\alpha$ is preferred since H$_\beta$ shows interference from other spectral features. Empirical formulae [25] are utilized for n_e determination from H$_\alpha$.

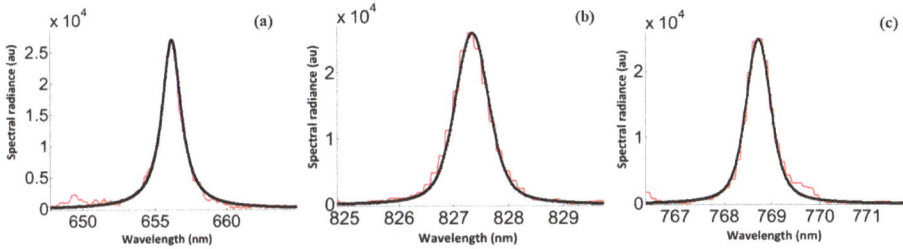

Figure 3. Fitting of line shapes (red) to Voigt line shape (black) for the H$_\alpha$-line (**a**) and the Ag I lines at 827.35 (**b**) and 768.7 nm (**c**) at fixed laser fluence 9.6 J/cm^2 and IR laser at 1064 nm excitation.

Table 1. Electron densities inferred from different spectral lines.

Laser Fluence (J/cm^2)	n_e (H$_\alpha$) (10^{17} cm^{-3})	$n_e^{827.35}$ Ag I: 827.35 nm (10^{17} cm^{-3})	$n_e^{768.9}$ Ag I: 768.7 nm (10^{17} cm^{-3})
9.9	1.64	1.66	1.76
7.5	0.76	0.77	0.76
5.9	0.63	0.66	0.7
4.5	0.57	0.55	0.58

The results in Table 1 attest that the two lines at wavelengths of 768.7 nm and 827.35 nm can be utilized for reliable measurements of plasma electron density at the surface of silver nano-based targets during the interaction with the blue Nd:YAG laser radiation.

This work also explores optical depths of plasma created by a laser at the green and blue wavelengths. In general, the laser produced plasma is inhomogeneous even though the laser device operates in the lowest order transverse electromagnetic mode, i.e., TEM$_{00}$ [1–10]. Actually, there are two regions in the plasma produced by laser radiation. The first region includes the central hot core with a relatively large electron temperature, density, and large population densities of higher emitting species. The second region includes the outer periphery region at which the plasma becomes relatively cold (losses of internal energy by adiabatic expansion against surrounding medium) [26,27]. The second region contains a large population in lower excitation states. This situation enhances the chance or probability for some generated photons at the central region to be re-absorbed by the cold atomic species at the peripheries [26,27].

The re-absorption processes act differently over the spectral line shape. The effect at the central upshifted spectral line with little effect at the line wings is called self-reversal [28–30] and it produces a dip, which is indicated in Figure 4. The other re-absorption process is self-absorption. It is difficult to assess the level of self-absorption in n$_e$ determinations but usually larger electron densities are found from self-absorbed lines. Comparisons with well-established, optically thin emission lines can be utilized to evaluate the level of self-absorption. The existence of certain, reliable, optically thin lines can provide a measure from which one can deduce if a recorded line is optically thin or thick.

Therefore, the standard H_α-line or other optically thin lines like the Ag I lines at 768.7 nm or 827.35 nm become important since one can use either line to check the value of the electron density.

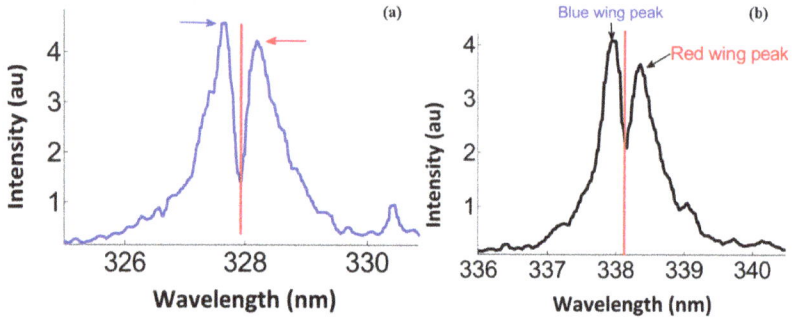

Figure 4. Spectral dips and asymmetries at a fluence of 9.6 J/cm^2. (**a**) 327.9 nm and (**b**) 338.2 nm.

Comparisons of the electron densities inferred from questionable lines with n_e obtained from optically thin lines can yield the self-absorption parameter, *SA*, by using the following formula $SA \simeq \left(\frac{1-\exp(-\tau_{SA})}{\tau_{SA}}\right)$ [23]. In this scenario, $\tau_{SA} \simeq \int_{-\ell}^{0} \kappa_{SA}(\lambda)\ d\lambda$ is the plasma optical depth due to the self-absorption coefficient $\kappa_{SA}(\lambda)$ integrated over the whole spectral line region, $\Delta\lambda$, along with the line-of-sight for plasma having an approximate length ℓ. Self-absorption causes line-shape distortions including the appearance of broader profiles with smaller spectral radiances than for optically thin lines. It is pointed out in Reference [31] that a correction to a spectral line shape against self-absorption can be carried out using a practical relation, $SA \simeq \left(n_e^{line}/n_e^*\right)^{-1.785}$, where n_e^* is the electron density deduced from largely optically thin lines, e.g., H_α at 656.3 nm, Ag I at 768.7 or 827.35 nm, which is mentioned before. As the *SA* parameter approaches unity, the line can be considered as optically thin. In other words, the *SA* parameter determines the degree of the plasma opacity for selected spectral lines.

Similarly, the self-reversal effect can be quantified with coefficient $SR \simeq \left(\frac{1-\exp(-\tau_{SR})}{\tau_{SR}}\right)$ that can be related to transmittance $T(\tau_{SR}) \approx \left(1/\sqrt{\pi\tau_{SR}}\right)$ [28–30] for plasma of optical depth τ_{SR}. The *SR* parameter is typically not much smaller than one, but it can cause a characteristic dip at the center of the line. However, *SR* has little effect in the line wings [30]. For example, Figure 4a,b display dips for the resonance lines Ag I at 327.9 nm and 338.2 nm.

Figure 4a,b display self-reversed Ag I resonance lines at wavelengths of 327.9 nm and 338.2 nm. The spectra are recorded when using 355-nm wavelength and 9.6 J/cm^2 fluence excitation. Two major features occur: (i) the dip at each of the lines centers and (ii) the pronounced line asymmetry of both lines.

Notice the asymmetries in the line shapes that appear to affect both lines. The theory that can describe the spectral dip due to self-reversal does not contain an asymmetry. However, the apparent asymmetry will be subjected to further investigation in a separate study. Figure 5 displays spectral line shapes of the 327.9-nm resonance line upon 355-nm irradiation for varying levels of fluence.

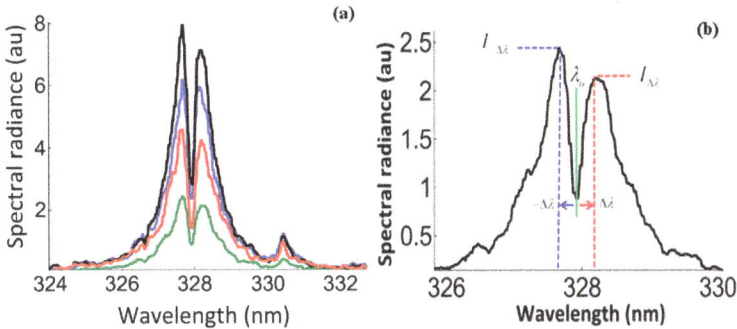

Figure 5. (a) Spectral radiance of the 327.9-nm line for different laser fluence. (b) Details for measurement of the spectral asymmetry.

The observed spectral line asymmetry, As, that appears for the 327.9-nm line can be calculated with $As = 2\left(\frac{I_{\Delta\lambda}-I_{-\Delta\lambda}}{I_{\Delta\lambda}+I_{-\Delta\lambda}}\right) \approx \left(\frac{hc}{\lambda_o k T_R} - 6\right)\frac{2\Delta\lambda}{\lambda_o}$ [4]. In this case, $I_{\Delta\lambda}$ and $I_{-\Delta\lambda}$ are the peak spectral radiances at red and blue wings, which is shown in Figure 5b. T_R denotes the radiation temperature not the electron temperature T_e because the laser-induced plasma is in local thermodynamic equilibrium (LTE), which implies that these temperatures are not equal i.e., $T_R \neq T_e$. Table 2 shows the amount of line asymmetry for decreasing laser fluence as well as values for the self-absorption factor, SA, and the Keldysh parameter, γ. The Keldysh parameter [32] is a measurement for the excitation mechanism, the $\gamma << 1$ and $\gamma >> 1$ imply tunnel, and multiphoton ionization, respectively. Table 2 shows that multiphoton ionization generates laser-induced plasma in spite of large electromagnetic fields that may occur near nanoparticles.

Table 2. Line asymmetries, self-absorption factors, and Keldysh parameters for different fluence.

Fluence (J/cm²)	13.4	10.9	6.9	5.6	4.8	3.9	2.1
Line Asymmetry	0.04	0.057	0.12	0.14	0.071	0.087	0.11
Self-absorption, SA	0.01	0.011	0.007	0.0048	0.0044	0.0022	0.0021
Keldysh parameter, γ	470	520	660	730	790	880	1200

5. Discussion

Recent work elaborates on the interaction of the Nd: YAG radiation at wavelengths of 1064 nm, 532 nm, and 355 nm with silver nano-based targets and for different laser fluence levels in the range of 2 J/cm² to 13 J/cm². Three observations are identified for the resonance lines at 327.9 nm and 338.2 nm. Investigations of the spectral lines shapes reveal that:

(I)   Self-reversal is characterized by a large dip at the central wavelength [33],
(II) Self-absorption, and
(III) Asymmetries.

In the previously reported investigations, there was minimal if any experimentally recognizable trends of line asymmetry variations with either laser fluence or electron density. Yet for neutral emitters, asymmetric line profiles due to the Stark effect are predicted [4]. However, for further explanations of the observed phenomena in future work, one should evaluate effects associated with internally generated micro electromagnetic fields [4–7,24,33] in nanomaterial.

In addition, the spectral lines that arise from the Ag I at wavelengths of 768.7 nm and 827.35 nm are found to be optically thin. Consistent results are found when comparing electron densities to those obtained from H_α. Consequently, these two Ag I lines can be used as a standard spectral line to determine the plasma electron density when H_α is absent in the measured spectrum.

6. Conclusions

The emitted resonance spectral lines in nano-enhanced laser-induced plasma spectroscopy indicate self-reversal and asymmetries. Internal nanomaterial electromagnetic fields in the plasma may affect the observed line asymmetry. Further experimental and theoretical efforts are recommended for the explanation of the spectral line shapes from the nanomaterial plasma. Several lines at near-IR wavelengths are optically thin and these lines can be used as a reliable indicator of plasma electron density.

Author Contributions: A.M.EL.S. designed and performed the experiments; A.M.EL.S. and A.E.EL.S. analyzed the result together with C.G.P. and all three authors contributed to the writing of the article.

Conflicts of Interest: The authors declare no conflict of interest.

References

1. Cremers, D.A.; Radziemski, L.J. *Handbook of Laser-Induced Breakdown Spectroscopy*; Wiley: Hoboken, NJ, USA, 2006; ISBN 978-0-470-09299-6.
2. Cremers, D.A.; Radziemski, L.J. *Laser Spectroscopy and its Applications*; Solarz, R.W., Paisner, J.A., Eds.; Marcel Dekker: New York, NY, USA, 1987; ISBN 978-0824775254.
3. Parigger, C.G. Laser-induced breakdown in gases: Experiments and simulation. In *Laser-Induced Breakdown Spectroscopy*; Miziolek, A.W., Palleschi, V., Schechter, I., Eds.; Cambridge University Press: Cambridge, UK, 2006; ISBN 978-0521852746.
4. Griem, H.R. *Plasma Spectroscopy*; McGraw-Hill Book Company: New York, NY, USA, 1964; LCCN 63023250.
5. Fujimoto, T. *Plasma Spectroscopy*; Clarendon Press: Oxford, UK, 2004; ISBN 9780198530282.
6. Kunze, H.-J. *Introduction to Plasma Spectroscopy*; Springer: New York, NY, USA, 2009; ISBN 978-3-642-02232-6.
7. Holtgreven, W.L. *Plasma Diagnostics*; Wiley: Amsterdam, The Netherlands, 1968; ISBN 978-0720401370.
8. EL Sherbini, A.M.; Aboulfotouh, A.M.; Parigger, C.G. Electron number density measurements using laser-induced breakdown spectroscopy of ionized nitrogen spectral lines. *Spectrochim. Acta Part B At. Spectrosc.* **2016**, *125*, 152–158. [CrossRef]
9. Oks, E. *Stark Broadening of Hydrogen and Hydrogen-Like Spectral Lines on Plasmas: The Physical Insight*; Alpha Science International: Oxford, UK, 2006; ISBN 1-84265-252-4.
10. Parigger, C.G.; Plemmons, D.H.; Oks, E. Balmer series Hβ measurements in a laser-induced hydrogen plasma. *Appl. Opt.* **2003**, *42*, 5992–6000. [CrossRef] [PubMed]
11. Oks, E. The shape of spectral lines of two-electron rydberg atoms/ions: Analytical solution. *J. Phys. B At. Mol. Opt. Phys.* **2017**, *50*, 115001. [CrossRef]
12. Pokropivny, V.; Lohmus, R.; Hussainova, I.; Pokropivny, A.; Vlassov, S. *Introduction in Nanomaterials and Nanotechnology*; University of Tartu, Tartu Press: Tartu, FI, USA, 2007; ISBN 978-9949-11-741-3.
13. Marambio-Jones, C.; Hoek, E.M.V. A review of the antibacterial effects of silver nanomaterials and potential implications for human health and the environment. *J. Nanopart. Res.* **2010**, *12*, 1531–1551. [CrossRef]
14. Campillo, I.; Guerrero, A.; Dolado, J.S.; Porro, A.; Ibáñez, S.; Goñi, A. Improvement of initial mechanical strength by nanoalumina in belite cements. *Mater. Lett.* **2007**, *61*, 1889–1892. [CrossRef]
15. Yurkov, G.Y.; Fionov, A.S.; Koksharov, Y.A.; Koleso, V.V.; Gubin, S.P. Electrical and magnetic properties of nanomaterials containing iron or cobalt nanoparticles. *Inorg. Mater.* **2007**, *43*, 834–844. [CrossRef]
16. Ray, P.C. Size and shape dependent second order nonlinear optical properties of nanomaterials and their application in biological and chemical sensing. *Chem. Rev.* **2010**, *110*, 5332–5365. [CrossRef] [PubMed]
17. Yang, H.; Liu, C.; Yang, D.; Zhang, H.; Xi, Z. Comparative study of cytotoxicity, oxidative stress and genotoxicity induced by four typical nanomaterials: The role of particle size, shape and composition. *J. Appl. Toxicol.* **2009**, *29*, 69–78. [CrossRef] [PubMed]
18. Ohta, T.; Ito, M.; Kotani, T.; Hattoti, T. Emission enhancement of laser-induced breakdown spectroscopy by localized surface plasmon resonance for analyzing plant nutrients. *Appl. Spectrosc.* **2009**, *63*, 555–558. [CrossRef] [PubMed]

19. EL Sherbini, A.M.; Aboulfotouh, A.-N.M.; Rashid, F.; Allam, S.; EL Dakrouri, T.A. Observed enhancement in LIBS signals from Nano vs. Bulk ZnO targets: Comparative study of plasma parameters. *World J. Nanosci. Eng.* **2012**, *2*, 181–188. [CrossRef]

20. EL Sherbini, A.M.; Galil, A.A.; Allam, S.H.; EL Sherbini, T.M. Nanomaterials induced plasma spectroscopy. *J. Phys. Conf. Ser.* **2014**, *548*, 012031. [CrossRef]

21. EL Sherbini, A.M.; Parigger, C.G. Wavelength dependency and threshold measurements for nanoparticle-enhanced laser-induced breakdown spectroscopy. *Spectrochim. Acta Part B At. Spectrosc.* **2016**, *116*, 8–15. [CrossRef]

22. EL Sherbini, A.M.; Parigger, C.G. Nano-material size dependent laser-plasma thresholds. *Spectrochim. Acta Part B At. Spectrosc.* **2016**, *124*, 79–81. [CrossRef]

23. El Sherbini, A.M.; El Sherbini, T.M.; Hegazy, H.; Cristoforetti, G.; Legnaioli, S.; Palleschi, V.; Pardini, L.; Salvetti, A.; Tognoni, E. Evaluation of self-absorption coefficients of aluminum emission lines in laser-induced breakdown spectroscopy measurements. *Spectrochim. Acta Part B At. Spectrosc.* **2005**, *60*, 1573–1579. [CrossRef]

24. Dimitrijević, M.S.; Sahal-Bréchot, S. Stark broadening of Ag I spectral lines. *At. Data Nucl. Data Tables* **2003**, *85*, 269–290. [CrossRef]

25. Surmick, D.M.; Parigger, C.G. Empirical formulae for electron density diagnostics from H_α and H_β line profiles. *Int. Rev. At. Mol. Phys.* **2014**, *5*, 73–81.

26. Mulser, P.; Bauer, D. *High Power Laser–Matter Interaction*; Springer: Heidelberg, Germany, 2010; ISBN 978-3-540-46065-7.

27. Hahn, D.; Omenetto, N. Laser-induced breakdown spectroscopy (LIBS), part II: Review of instrumental and methodological approaches to material analysis and applications to different fields. *Appl. Spectrosc.* **2012**, *66*, 347–419. [CrossRef] [PubMed]

28. Holstein, T. Imprisonment of resonance radiation in gases. *Phys. Rev.* **1947**, *72*, 1212–1233. [CrossRef]

29. Holstein, T. Imprisonment of resonance radiation in gases. II. *Phys. Rev.* **1951**, *83*, 1159–1168. [CrossRef]

30. Irons, F.E. The escape factor in plasma spectroscopy—I. The escape factor defined and evaluated. *J. Quant. Spectrosc. Radiat. Transf.* **1979**, *22*, 1–20. [CrossRef]

31. Konjević, N. Plasma broadening and shifting of non-hydrogenic spectral lines: Present states and applications. *Phys. Rep.* **1999**, *316*, 339–401. [CrossRef]

32. Keldysh, L.V. Ionization in the field of a strong electromagnetic wave. *Sov. Phys. J. Exp. Theor. Phys.* **1965**, *20*, 1307–1314.

33. Hey, J.D.; Korten, M.; Lie, Y.; Pospieszczyk, T.; Rusbüldt, A.; Schweer, B.; Unterberg, B.; Wienbeck, J.; Hintz, E. Doppler broadening and magnetic field effects on the balmer lines emitted at the edge of a tokamak plasma. *Contrib. Plasma Phys.* **1996**, *36*, 583–604. [CrossRef]

![atoms logo] *atoms*

MDPI

Article

Improving the Method of Measuring the Electron Density via the Asymmetry of Hydrogenic Spectral Lines in Plasmas by Allowing for Penetrating Ions

Paul Sanders and Eugene Oks *

Physics Department, 206 Allison Lab, Auburn University, Auburn, AL 36849, USA; phs0007@auburn.edu
* Correspondence: goks@physics.auburn.edu

Received: 9 March 2018; Accepted: 8 April 2018; Published: 18 April 2018

Abstract: There was previously proposed and experimentally implemented a new diagnostic method for measuring the electron density N_e using the asymmetry of hydrogenic spectral lines in dense plasmas. Compared to the traditional method of deducing N_e from the experimental widths of spectral lines, the new method has the following advantages. First, the traditional method requires measuring widths of at least two spectral lines (to isolate the Stark broadening from competing broadening mechanisms), while for the new diagnostic method it is sufficient to obtain the experimental profile of just one spectral line. Second, the traditional method would be difficult to implement if the center of the spectral lines was optically thick, while the new diagnostic method could still be used even in this case. In the theory underlying this new diagnostic method, the contribution of plasma ions to the spectral line asymmetry was calculated only for configurations where the perturbing ions were outside the bound electron cloud of the radiating atom/ion (non-penetrating configurations). In the present paper, we take into account the contribution to the spectral line asymmetry from *penetrating configurations*, where the perturbing ion is inside the bound electron cloud of the radiating atom/ion. We show that in high-density plasmas, the allowance for penetrating ions can result in significant corrections to the electron density deduced from the spectral line asymmetry.

Keywords: asymmetry of spectral lines; penetrating ions; spectroscopic diagnostics of plasmas; electron density measurements

---

1. Introduction

In medium-density plasmas, profiles of hydrogenic spectral lines look symmetric, but in high-density plasmas, they become asymmetric. This asymmetry is caused primarily by the nonuniformity of the ion microfield, as noted by Sholin and his co-workers in papers [1–3]—for the latest advances in the theory of the asymmetry we refer to papers [4,5] and the references therein, of which we especially note papers [6,7]. (There are also secondary sources of the asymmetry, as discussed in more detail below in the first paragraph of Section 2). Often, the blue maximum of the spectral line is higher than the red maximum, and the positions of the intensity maxima are asymmetrical with respect to the unperturbed line center.

A new diagnostic method for measuring the electron density using the asymmetry of hydrogenic spectral lines in dense plasmas was proposed and implemented in paper [8]. In that paper, in particular, from the experimental asymmetry of the C VI Lyman-delta line emitted by a vacuum spark discharge, the electron density was deduced to be $N_e = 3 \times 10^{20}$ cm^{-3}. This value of N_e was in good agreement with the electron density determined from the experimental widths of C VI Lyman-beta and Lyman-delta lines.

Later, this diagnostic method was also employed in the experiment presented in paper [9]. In that laser-induced breakdown spectroscopy experiment, the electron density $N_e \sim 3 \times 10^{17}$ cm^{-3} was determined from the experimental asymmetry of the H I Balmer-beta (H-beta) line.

This new diagnostic method has the following advantages compared to the method of deducing N_e from the experimental widths of spectral lines. First, the latter, traditional, method requires measuring widths of at least two spectral lines—because the widths are affected not only by the Stark broadening, but also by competing broadening mechanisms, such as, e.g., Doppler broadening. In distinction, to use the new diagnostic method, it is sufficient to obtain the experimental profile of just one spectral line—because the Doppler broadening does not cause the asymmetry.

Second, the traditional method based on experimental widths would be difficult to implement if the center of the spectral lines were optically thick. In distinction, the new diagnostic method can still be used even if the spectral line is optically thick in its central part. This is because the overwhelming contribution to the asymmetry originates from the wings of the spectral line, the wings usually being optically thin. More details can be found in Section 1.6 of [10][1].

In the theory underlying this new diagnostic method, the contribution of plasma ions to the spectral line asymmetry was calculated only for configurations where the perturbing ions were outside the "atomic sphere", i.e., outside the bound electron cloud of the radiating atom/ion (non-penetrating configurations). In the present paper, we take into the contribution to the spectral line asymmetry from *penetrating configurations*, i.e., from configurations where the perturbing ion is inside the bound electron cloud of the radiating atom/ion (hereafter, radiator). We show that, in high-density plasmas, the allowance for penetrating ions can result in significant corrections to the electron density deduced from the spectral line asymmetry.

2. Allowance for Penetrating Ions

Let us first present a brief overview of the underlying theory for non-penetrating configurations. The dipole interaction of the radiator with perturbing ions outside the bound electron cloud, being calculated in the first order of the perturbation theory, splits the spectral line into Stark components symmetrically with respect to the unperturbed frequency or wavelength—in terms of both positions and intensities of the Stark components. The quadrupole interactions of the radiator with perturbing ions outside the bound electron cloud, being calculated in its first nonvanishing order, causes the asymmetry of the Stark splitting—in terms of both positions and intensities of the Stark components. The latter is the primary source of the asymmetry: other sources of the asymmetry—such as, but not limited to, e.g., the dipole interaction in the second order (known as the quadratic Stark effect), the quadrupole interaction in the second order, the octupole interaction in the first order—add to the asymmetry only higher-order corrections in terms of the corresponding small parameter $n^2 a_0 N_e^{1/3}/Z^{4/3}$, where n is the principal quantum number of the upper level involved in the radiative transition, a_0 is the Bohr radius, Z is the charge of plasma ions and the nuclear charge of the radiating ion.[2]

---

[1] We note that Ref. [40] from chp. 1 of [10] on the paper referred to here as [8] has typographic errors. The correct one is our Ref. [8] here.

[2] The Boltzmann factor $\exp(-\hbar \, \Delta\omega/T)$ also contributes to the asymmetry (here $\Delta\omega$ is the detuning from the unperturbed frequency of the spectral line and T is the temperature). For quasistatic wings, $\hbar \, \Delta\omega/T$ scales with the electron density as $(a_0 N_e^{1/3})^2$. Therefore, for plasmas of the electron densities $N_e \ll 6.7 - 10^{24}$ cm^{-3} (the right side being the atomic unit of the electron density), the Boltzmann factor contribution to the asymmetry is much smaller than the quadrupole interaction contribution to the asymmetry that scales as $a_0 N_e^{1/3}$. Additionally, there is also the factor $(1 + \omega/\omega_0)^4$ caused by the scaling of the dipole radiation intensity (ω_0 being the unperturbed frequency). The asymmetry contributions of this factor and of the Boltzmann factor essentially cancel each other out (see, e.g., Section 5.11 of paper [5]). (Continued at the bottom of the next page). There is also so-called trivial contribution to the asymmetry caused by the conversion from the frequency scale to the wavelength scale. This consists of two factors (see, e.g., Section 5.8 of paper [5]): the transformation of the argument $\Delta\omega$ (given by Equation (24) from [5]) and the transformation of the intensity (given by Equation (25) from [5]). These two factors essentially cancel each other out (as shown in [5]), so that the resulting trivial contribution to the asymmetry is much smaller than the quadrupole interaction contribution to the asymmetry.

However, in paper [11], it was shown that the quadrupole interaction, despite causing the asymmetric splitting of the spectral line into Stark components, does not shift the center of gravity of the line profile. Therefore, in the new diagnostic method presented in paper [8], first the center of gravity of the experimental profile was determined, and then it was taken as the reference point. Then, with respect to this point, the integrated intensities of the blue (I_B) and red (I_R) wings of the experimental profile were found. After that, the experimental degree of asymmetry, defined as

$$\rho_{quad} = \frac{I_B - I_R}{0.5[I_B + I_R]},$$ (1)

was determined and then compared with the corresponding theoretical value given below.

The theoretical intensities of the blue and red wings, resulting from dipole and quadrupole interactions of the radiator with perturbing ions outside the bound electron cloud, can be expressed as follows (see paper [8]):

$$I_B = \sum_{k>0} I_k^{(0)} \left(1 + \frac{a_o}{Z_r R_o} \epsilon_k^{(1)} \langle \frac{R_0}{R} \rangle \right),$$ (2)

and

$$I_R = \sum_{k<0} I_k^{(0)} \left(1 + \frac{a_o}{Z_r R_o} \epsilon_k^{(1)} \langle \frac{R_0}{R} \rangle \right),$$ (3)

where Z_p is the charge of perturbing ions, Z_r is the nuclear charge of the radiator, a_o is the Bohr radius, and $R_o = [(4\pi/3)N_p]^{-1/3}$ is the mean interionic distance, $N_p = N_e/Z_p$ being the perturbing ion density. Here, $I_k^{(0)}$ and $\epsilon_k^{(1)}$ are the unperturbed intensity and the quadrupole correction to the intensity, respectively, the subscript k being the label of Stark components of the spectral line; $k > 0$ and $k < 0$ correspond to the blue-shifted and red-shifted components, respectively (the values of $I_k^{(0)}$ and $\epsilon_k^{(1)}$ for several Lyman and Balmer lines were tabulated in paper [2]). The quantity $\langle R_0/R \rangle$ is the scaled inverse distance between the perturbing ion and the radiator averaged over the distribution of such distances.

Finally, the theoretical degree of asymmetry was presented in paper [8] in the form:

$$\rho_{quad} = 0.46204 \left(\frac{N_e \ [cm^{-3}]}{10^{21}} \right)^{\frac{1}{3}} \frac{1}{Z_p^{\frac{1}{3}} Z_r} \sum_{k>0} I_k^{(0)} \epsilon_k^{(1)},$$ (4)

Then the electron density N_e was determined in paper [8] by substituting the experimental degree of asymmetry into the left side of Equation (4).

In the present paper, we consider the contribution of penetrating ions to the spectral line asymmetry in order to refine this diagnostic method. For simplicity, we limit ourselves below to the practically important case $Z_p = Z_r = Z$. The energy shifts due to penetrating ions can be calculated by the perturbation theory on the basis of the spherical wave functions of the so-called "united atom" of the nuclear charge $2Z$.

The perturbed energy shifts (counted from the unperturbed energies) for the orbital quantum number $l > 0$ are given by (see, e.g., Equations (6) and (7) from paper [12] or Equations (5.11) and (5.12) from book [13]):

$$E_{nlm} = -\frac{8 \left[l(l+1) - 3m^2 \right] Z^4 R^2 \ e^2}{a_o^3 \ n^3 l(l+1)(2l-1)(2l+1)(2l+3)}.$$ (5)

For the case of $l = 0$, the calculated energy shift is:

$$E_{n00} = \frac{8 \ Z^4 R^2 e^2}{3 \ a_o^3 n^3}.$$ (6)

We note that Equation (6) can also be obtained from Equation (5), first by setting m = 0, and then by canceling out $l(l + 1)$ in the numerator and denominator, and by setting $l = 0$. (This was mentioned in book [12], but in Equation (5.11) from [12] corresponding to our Equation (6), there was a typographic error in the sign.)

The frequency change of an individual Stark component is thus given by

$$\Delta\omega_k = -\frac{Z^2 e^2 \Delta_k^1}{2 \hbar a_o^3} R^2, \tag{7}$$

where

$$\Delta_k^1 = 16 Z^2 \left[\frac{l(l+1) - 3 m^2}{n^3 l(l+1)(2l-1)(2l+1)(2l+3)} - \frac{l'(l'+1) - 3 m'^2}{n'^3 l'(l'+1)(2l'-1)(2l'+1)(2l'+3)} \right]. \tag{8}$$

For the specific case where either $l = 0$ or $l' = 0$, Equation (8) reduces to

$$\Delta_k^1 = \begin{cases} 16 Z^2 \left[\frac{1}{3 n^3} - \frac{l'(l'+1) - 3 m'^2}{n'^3 l'(l'+1)(2l'-1)(2l'+1)(2l'+3)} \right], l = 0; l' \neq 0 \\ 16 Z^2 \left[\frac{l(l+1) - 3 m^2}{n^3 l(l+1)(2l-1)(2l+1)(2l+3)} - \frac{1}{3 n'^3} \right], l' = 0; l \neq 0 \end{cases}. \tag{9}$$

Then, the quasi-static profile of each Stark component can be represented in the form:

$$S_k(\Delta\lambda) = \int_0^{u_{max}} W(u) \left[I_k^{(0)} + I_k^{(1)} \right] \delta \left(\Delta\lambda - \frac{Z^2 e^2 \Delta_k^1 \lambda_0^2}{4 \pi c \hbar a_o^3} u \right) du. \tag{10}$$

Here, $u \equiv R^2$, and the probability of finding the perturbing ion a distance u away from the radiating atom is taken to be the binary distribution. For simplifying the integration, we use the expansion of the distribution in powers u/R_0^2 and keep the terms up to $\sim u^2$:

$$W(u)du = \frac{3 \sqrt{u}}{2 R_0^3} \exp\left(-\frac{\sqrt{u}^3}{R_0^3}\right) du \approx \frac{3 \sqrt{u}}{2 R_0^3} - \frac{3 u^2}{2 R_0^6}. \tag{11}$$

For the case of a hydrogenic radiator under the presence of a penetrating ion, the relative intensities of each line component can be best calculated analytically using the robust perturbation theory developed by Oks and Uzer [14]. A more detailed explanation of this procedure is outlined in Appendix A. The relative intensities of each component can be written as

$$I_k = \Delta_{lk}^0 + Z^2 \Delta_{lk}^1 u^2, \tag{12}$$

where Δ_{lk}^0 and Δ_{lk}^1 are tabulated in Appendix B for each component of the spectral line Balmer-alpha, considered here as an example. These coefficients represent corrections to the intensity of the line, calculated from the perturbation theory briefly mentioned above.

The upper limit u_{max} of the integration in Equation (10) should be the smallest of the following two "candidates". One candidate for u_{max} is the root mean square size of the bound electron cloud, which depends on the sublevel in consideration:

$$r_{rms} = \sqrt{\frac{n^2}{2 Z^2} [5 n^2 + 1 - 3 l (l + 1)]}. \tag{13}$$

The other candidate for u_{max} is defined by the limit of the applicability of the perturbation theory. Of course, this would ensure that formally calculated corrections to the energy and intensities of the spectral line would remain relatively small.

The allowance for penetrating ions shifts the center of gravity of the spectral line, as shown in paper [15]. (This is the only contribution to the shift of the center of gravity, since the dipole and quadrupole interactions of the radiator with perturbing ions outside the bound electron cloud do not shift the center of gravity, as shown in paper [11] and mentioned above). For the He II Balmer-alpha line, which we use as an example, the center of gravity shift due to penetrating ions was calculated analytically in paper [15] to be

$$\Delta \lambda_{PI}\left(m\text{Å}\right) = 17 \frac{N_e\left(cm^{-3}\right)}{10^{17}}. \tag{14}$$

So, with the allowance for penetrating ions, the reference point for calculating the integrated intensities of the blue and red wings must be shifted by the amount given by Equation (14).

After carrying out the integration in Equation (10), the profile reduces to

$$S_k(\Delta \lambda) = \left(\frac{Z^2 e^2 \Delta_k^1 \lambda_0^2}{4\pi c \hbar a_o^3}\right)^{-1} I_k(u_0) \left(\frac{3u_0^{\frac{1}{2}}}{2 R_o^3} - \frac{3u_0^2}{2 R_o^6}\right) \Theta\left[u_{max} \frac{Z^2 e^2 \Delta_k^1 \lambda_0^2}{4\pi c \hbar a_o^3} - |\Delta \lambda|\right], \tag{15}$$

where $\Theta[\ldots]$ is the Heaviside step function and u_0 is the root of the delta function, given by

$$u_0 = \frac{4\pi c \hbar a_o^3}{Z^2 e^2 \Delta_k^1 \lambda_0^2} \Delta \lambda. \tag{16}$$

Thus, for the contributions of the penetrating ions to the integrated intensities of the blue and read parts of the line profile, we get

$$I_{PI,B} = \sum_{k<0} \int_{-\Delta \lambda_{max}}^{-\Delta \lambda_{PI}} S_k(\Delta \lambda) d\Delta \lambda \tag{17}$$

and

$$I_{PI,R} = \sum_{k>0} \int_{\Delta \lambda_{PI}}^{\Delta \lambda_{max}} S_k(\Delta \lambda) d\Delta \lambda, \tag{18}$$

respectively. Here

$$\Delta \lambda_{max} = u_{max} \frac{Z^2 e^2 \Delta_k^1 \lambda_0^2}{4\pi c \hbar a_o^3}, \tag{19}$$

which is obtained by equating to zero the argument of the Heaviside step function. Additionally, what is meant in Equations (17) and (18) by $k < 0$ (or $k > 0$) is the inclusion of only those components which involve corrections to the energy that are positive (or negative), implying blue-shifted (or red-shifted) components of the spectral line.

By combining the above result with the contribution of the quadrupole interaction (the interaction of the radiator with perturbing ions outside the bound electron cloud) to the integrated intensities of the blue and read parts of the profile, we obtain our final result for the degree of asymmetry

$$\rho_{act} = \frac{I_B + I_{PI,B} - I_R - I_{PI,R}}{0.5[I_B + I_{PI,B} + I_R + I_{PI,R}]}, \tag{20}$$

where subscript *act* stands for *actual*—in distinction to ρ_{quad}.

The combination of Equations (4) and (20) connects the degree of asymmetry with the electron density N_e and thus allows a more accurate determination of the electron density from the experimental asymmetry. We illustrate this below by the example of the He II Balmer-alpha line.

Table 1 presents the following quantities for the He II Balmer-alpha line at five different values of the actual electron density:

-   the theoretical degree of asymmetry ρ_{act} calculated with the allowance for penetrating ions,
-   the theoretical degree of asymmetry ρ_{quad} calculated without the allowance for penetrating ions,

- the electron density $N_{e,quad}$ that would be deduced from the experimental asymmetry degree while disregarding the contribution of the penetrating ions,
- the relative error $|N_{e,quad} - N_{e,act}|/N_{e,act}$ in determining the electron density from the experimental asymmetry degree while disregarding the contribution of the penetrating ions.

Table 1. The relative error in determining the electron density N_e from the experimental asymmetry degree while disregarding the contribution of the penetrating ions for the He II Balmer-alpha line. The physical quantities in Table 1 are explained in the text directly above Table 1.

$N_{e,act}/(10^{18}$ cm$^{-3})$	ρ_{act}	ρ_{quad}	$N_{e,quad}/(10^{18}$ cm$^{-3})$	$|N_{e,quad} - N_{e,act}|/N_{e,act}$
2	0.0925	0.0955	1.82	9.03%
4	0.114	0.120	3.42	14.5%
6	0.128	0.138	4.86	19.1%
8	0.139	0.152	6.16	23.1%
10	0.147	0.163	7.33	26.7%

It is seen that in high-density plasmas, the allowance for penetrating ions can indeed result in significant corrections to the electron density deduced from the spectral line asymmetry.

3. Conclusions

To improve the diagnostic method for measuring the electron density using the asymmetry of spectral lines in dense plasmas, we took into account the contribution to the spectral line asymmetry from *penetrating configurations*, i.e., from the configurations where the perturbing ion is inside the bound electron cloud of the radiating atom/ion. After performing the corresponding analytical calculations, we demonstrated that in high-density plasmas, the allowance for penetrating ions can result in significant corrections to the electron density deduced from the spectral line asymmetry.

It is worth clarifying why we took into account the shift of the line *as a whole* due to penetrating ions, but did not take into account other mechanisms shifting the line *as a whole*, such as, e.g., plasma polarization shift and the shift by plasma electrons. The experimental integrated intensities of the blue (I_B) and red (I_R) parts of the profile are calculated with respect to the experimental center of gravity of the profile. The latter shifts of the line *as a whole* do not contribute to the asymmetry, and thus should not affect the experimental values of I_B and I_R. The reason why we took into account the shift of the line *as a whole* by penetrating ions is that penetrating ions contribute simultaneously to both the asymmetry and the shift of the line as the whole. Since these two effects of penetrating ions are two sides of the same coin, both of them should be taken into account.

We mention in passing that the potential transition of electrons into the quasistatic regime is practically irrelevant to the asymmetry. Indeed, as Demura and Sholin wrote in paper [3], for the quadrupole interaction $U \sim Q/R^3$, which controls the asymmetry of hydrogenic spectral lines, electrons can become quasistatic at the frequency detuning from the line center $\Delta\omega \sim v_{Te}^{3/2}/Q^{1/2}$, where v_{Te} is the mean thermal velocity of plasma electrons—according to Holstein [16] (see also Sobelman book [17]). However, such detuning significantly exceeds the mean separation between spectral lines—both for the Lyman and Balmer series, as noted by Demura and Sholin [3], thus making the potential transition of electron into the quasistatic regime irrelevant to the problem of asymmetry of hydrogenic spectral lines. A similar conclusion was drawn also in paper [18].

Finally, we note that the electron densities $N_e \sim (10^{18}-10^{19})$ cm^{-3}, which we used in the illustrative example of the He II Balmer-alpha line, are achievable in plasma spectroscopy. Examples include experiment [19], with a hydrogen plasma, and experiment [20], with a helium plasma.

Author Contributions: Both authors contributed equally to this work.

Conflicts of Interest: The authors declare no conflict of interests.

Appendix A. Details of Calculating Perturbed Matrix Elements

The redistribution of intensities of Stark components, along with wavelength shifts due to the presence of perturbing ions, play a crucial role in determining the degree of asymmetry of the spectral line. These values have been tabulated according to the robust perturbation theory developed by Oks and Uzer [14] based on using the super-generalized Runge-Lenz vector derived by Kryukov and Oks [21]. Since the unperturbed system has an additional constant of the motion (namely the Runge-Lenz vector), the task of calculating the corrections to the state is simplified. The reason for this beneficial result is that the correction to the Runge-Lenz vector is non-degenerate with respect to the same states that are degenerate in the correction to the Hamiltonian. The mixing of the states is elucidated by the Runge-Lenz vector correction under the influence of the perturbing ion. Here are some details, formulas being presented in atomic units.

According to paper [21], for the problem of an electron in the field of two Coulomb centers of charges Z_1 and Z_2, the additional conserved quantity is the following projection of the super-generalized Runge-Lenz vector on the internuclear axis

$$A_z = \mathbf{p} \times \mathbf{L} \cdot \mathbf{e_z} - L^2/R - Z_1 z/r - Z_2(R - z)/|\mathbf{R} - \mathbf{r}| + Z_2, \tag{A1}$$

where $\mathbf{p}$, $\mathbf{L}$, and $\mathbf{r}$ are the linear momentum, the angular momentum, and the radius-vector of the electron, respectively; $\mathbf{R}$ is the vector directed from charge Z_1 to charge Z_2. For the case where $R \ll r$, the unperturbed part A_{z0} of the operator A_z can be chosen as

$$A_{z0} = -L^2/R, \tag{A2}$$

corresponding to the unperturbed Hamiltonian of the so-called "united atom" of the nuclear charge $Z_1 + Z_2$:

$$H_0 = p^2/2 - (Z_1 + Z_2)/r. \tag{A3}$$

Operators H_0 and A_{z0} have common eigenfunctions (the spherical eigenfunctions of the Coulomb problem). The spectrum of eigenvalues of the operator H_0 is degenerate. . Therefore, calculating corrections to the eigenfunctions of the operator H_0 using the standard perturbation theory would require going to the 2nd order of the *degenerate* perturbation theory, thus involving generally infinite summations (see, e.g., the textbook [22]).

In distinction, the spectrum of eigenvalues of the operator A_{z0} is nondegenerate (the eigenvalues being—$l(l + 1)/R$). Therefore, the corrections to the eigenfunctions can be easily calculated in the 1st order of the standard *nondegenerate* perturbation theory. The coefficients of the corresponding linear combinations of the unperturbed eigenfunctions are

$$<nl'm|(A_z - A_{z0})|nlm>/\{[l'(l' + 1) - l(l + 1)]/R\} \tag{A4}$$

and do not involve infinite summations. This example is another illustration of the advantages of the robust perturbation theory developed in paper [14] over the standard perturbation theory.

In this way, by using the first nonvanishing term of the expansion of the operator $(A_z - A_{z0})$ in powers of R, we obtained the following expression for the 1st order corrections to the eigenfunctions for the specific case of $Z_1 = Z_2 = Z$:

$$\Psi_{nlm}^{(1)} = \frac{5\left[\frac{(l_>^2 - m^2)(n^2 - l_>^2)}{(2\,l_> + 1)(2\,l_> - 1)}\right]^{\frac{1}{2}}}{n\,[l(l+1) - l'(l' - 1)]}\,Z\,R\,\Psi_{nl'm'}^{(0)}, \tag{A5}$$

where $l_>$ denotes the greater value between l and l'. The selection rules are $l' = l \pm 1$ and $m' = m$.

We note that in the opposite case, where R >> r, the unperturbed part $A_{z1,0}$ of the operator A_z can be chosen in the usual way

$$A_{z1,0} = zp^2 - p_z(\mathbf{rp}) - Z_1 z/r, \tag{A6}$$

where the notation $(\mathbf{rp})$ stands for the scalar product (also known as the dot-product) of the operators $\mathbf{r}$ and $\mathbf{p}$. The corresponding unperturbed Hamiltonian is

$$H_{1,0} = p^2/2 - Z_1/r. \tag{A7}$$

The operator $A_{z1,0}$ has a nondegenerate spectrum of eigenvalues equal to q/n, where $q = (n_1 - n_2)$ is the difference of the parabolic quantum numbers. Therefore, the first nonvanishing corrections to the common eigenfunctions of the operators $H_{1,0}$ and $A_{z1,0}$ can be easily calculated in the 1st order of the standard *nondegenerate* perturbation theory. The coefficients of the corresponding linear combinations of the unperturbed eigenfunctions are

$$<nl'm|L^2|nlm>/[(q'/n - q/n)R], \tag{A8}$$

where $|q' - q| = 2$, as follows from the selection rules.

In distinction, to obtain the same corrections to the eigenfunctions using the operator $H_{1,0}$, whose spectrum of eigenvalues is degenerate, it would require going to the 2nd order of the *degenerate* perturbation theory and dealing with its complications, as Sholin did in his paper [2].

We note in passing that we also applied the robust perturbation theory [14] to analytically calculating corrections to the eigenfunctions due to the quadrupole interaction with ions *outside* the atomic electron cloud (non-penetrating ions), and we obtained the same analytical results as in the Sholin paper [2], but in a much simpler way. We also note that in paper [23], some corrections were presented to the input data from the tables in the Sholin paper [2].

Appendix B. Table of Intensities and Energy Level Corrections for the He II Balmer-alpha line

The perturbed intensity and frequency corrections for He II Balmer-alpha line are presented below. The quantum numbers of the upper and lower sublevels are in the spherical quantization.

Table A1. Corrections to the intensity and frequency of the He II Balmer-alpha line components.

Upper Sublevel	Lower Sublevel	Δ_{Ik}^0	Δ_{Ik}^1	Δ_k^1
322	211	$\frac{768}{4715}$	0	$\frac{173}{5670}$
321	211	$\frac{384}{4715}$	$-\frac{32}{14145}$	$\frac{197}{5670}$
321	210	$\frac{384}{4715}$	$-\frac{64}{2829}$	$-\frac{37}{567}$
321	200	0	$\frac{2792}{127305}$	$\frac{949}{2835}$
320	211	$\frac{128}{4715}$	$-\frac{128}{127305}$	$\frac{41}{1134}$
320	210	$\frac{512}{4715}$	$-\frac{3968}{127305}$	$-\frac{181}{2835}$
320	200	0	$\frac{11168}{381915}$	$\frac{953}{2835}$
311	211	0	$\frac{32}{14145}$	$\frac{19}{810}$
311	210	0	$\frac{232}{14145}$	$-\frac{31}{405}$
311	200	$\frac{160}{2829}$	$-\frac{400}{25461}$	$\frac{131}{405}$
310	211	0	$\frac{8}{3105}$	$\frac{43}{810}$
310	210	0	$\frac{2512}{127305}$	$-\frac{19}{405}$
310	200	$\frac{160}{2829}$	$-\frac{280}{8487}$	$\frac{143}{405}$
300	211	$\frac{5}{943}$	$\frac{40}{25461}$	$-\frac{53}{810}$
300	210	$\frac{5}{943}$	$-\frac{295}{101844}$	$-\frac{67}{405}$
300	200	0	$\frac{5525}{305532}$	$\frac{19}{81}$

References

1. Kudrin, L.P.; Sholin, G.V. Asymmetry of spectral lines of hydrogen in plasma. *Sov. Phys. Dokl.* **1963**, *7*, 1015.
2. Sholin, G.V. On the nature of the asymmetry of the spectral line profiles of hydrogen in a dense plasma. *Opt. Spectrosc.* **1969**, *26*, 275.
3. Demura, A.V.; Sholin, G.V. Theory of the asymmetry of hydrogen-line Stark profiles in a dense plasma. *J. Quant. Spectrosc. Radiat. Transf.* **1975**, *15*, 881–899.
4. Djurovic, S.; Ćirišan, M.; Demura, A.V.; Demchenko, G.V.; Nikolić, D.; Gigosos, M.A.; González, M.A. Measurements of H$_\beta$ Stark central asymmetry and its analysis through standard theory and computer simulations. *Phys. Rev. E* **2009**, *79*, 046402. [CrossRef] [PubMed]
5. Demura, A.V.; Demchenko, G.V.; Nikolic, D. Multiparametric dependence of hydrogen Stark profiles asymmetry. *Eur. Phys. J. D* **2008**, *46*, 111–127. [CrossRef]
6. Halenka, J.Z. Asymmetry of hydrogen lines in plasmas utilizing a statistical description of ion-quadruple interaction in Mozer-Baranger limit. *Physics D* **1990**, *16*, 1. [CrossRef]
7. Gunter, S.; Konies, A. Shifts and asymmetry parameters of hydrogen Balmer lines in dense plasmas. *Phys. Rev. E* **1997**, *55*, 907. [CrossRef]
8. Podder, N.K.; Clothiaux, E.J.; Oks, E. A method for density measurements employing an asymmetry of lineshapes in dense plasmas and its implementation in a vacuum spark discharge. *J. Quant. Spectrosc. Radiat. Transf.* **2000**, *65*, 441–453. [CrossRef]
9. Parigger, C.G.; Swafford, L.D.; Woods, A.C.; Surmick, D.M.; Witte, M.J. Asymmetric hydrogen beta electron density diagnostics of laser-induced plasma. *Spectrochim. Acta Part B* **2014**, *99*, 28–33. [CrossRef]
10. Oks, E. *Diagnostics of Laboratory and Astrophysical Plasmas Using Spectral Lineshapes of One-, Two-, and Three-Electron Systems*; World Scientific: Hackensack, NJ, USA; Singapore, 2017.
11. Oks, E. New type of shift of hydrogen and hydrogenlike spectral lines. *J. Quant. Spectrosc. Radiat. Transf.* **1997**, *58*, 821. [CrossRef]
12. Gavrilenko, V.P.; Oks, E.; Radchik, A.V. Hydrogen-like atom in the field of high-frequency linearly polarized electromagnetic radiation. *Opt. Spectrosc.* **1985**, *59*, 411.
13. Komarov, I.V.; Ponomarev, L.I.; Yu, S. *Slavyanov, Spheroidal and Coulomb Spheroidal Functions*; Nauka: Moscow, Russia, 1976. (In Russian)
14. Oks, E.; Uzer, T. A robust perturbation theory for degenerate states based on exact constants of the motion. *Europhys. Lett.* **2000**, *49*, 5. [CrossRef]
15. Sanders, P.; Oks, E. Estimate of the Stark Shift by Penetrating Ions within the Nearest Perturber Approximation for Hydrogenlike Spectral Lines in Plasmas. *J. Phys. B* **2017**, *50*, 245002. [CrossRef]
16. Holstein, T. Pressure broadening of spectral lines. *Phys. Rev.* **1950**, *79*, 744. [CrossRef]
17. Sobelman, I.I. *An Introduction to the Theory of Atomic Spectra*; Elsevier: Amsterdam, The Netherlands, 1972.
18. Oks, E.; Sholin, G.V. Boundary determination for electron quasistationarity using the asymmetry of the wings of hydrogen spectral line profiles. *Opt. Spectrosc.* **1972**, *33*, 217–218.
19. Kielkopf, J.F.; Allard, N.F. Shift and width of the Balmer series Hα line at high electron density in a laser-produced plasma. *J. Phys. B At. Mol. Opt. Phys.* **2014**, *47*, 155701. [CrossRef]
20. Marangos, J.P.; Burgess, D.D.; Baldwin, K.G.H. He II Balmer series line shifts in a dense z-pinch plasma. *J. Phys. B At. Mol. Opt. Phys.* **1988**, *21*, 3357. [CrossRef]
21. Kryukov, N.; Oks, E. Supergeneralized Runge-Lenz vector in the problem of two Coulomb or Newton centers. *Phys. Rev. A* **2012**, *85*, 054503. [CrossRef]
22. Landau, L.D.; Lifshitz, E.M. *Quantum Mechanics*; Pergamon: Oxford, UK, 1965.
23. Sanders, P.; Oks, E. Correcting the Input Data for Calculating the Asymmetry of Hydrogenic Spectral Lines in Plasmas. *Atoms* **2018**, *6*, 9. [CrossRef]

![atoms logo] *atoms*

MDPI

Article

Automodel Solutions of Biberman-Holstein Equation for Stark Broadening of Spectral Lines

Alexander B. Kukushkin [1,2,3,*], **Vladislav S. Neverov** [1], **Petr A. Sdvizhenskii** [1] and **Vladimir V. Voloshinov** [4]

[1] National Research Center "Kurchatov Institute", 123182 Moscow, Russia; vs-never@hotmail.com (V.S.N.); sdvinpt@gmail.com (P.A.S.)
[2] Institute of Laser and Plasma Technologies, National Research Nuclear University MEPhI, 115409 Moscow, Russia
[3] Institute of Nano-, Bio-, Information, Cognitive and Socio-Humanistic Sciences and Technologies, Moscow Institute Physics and Technology (State University), 141700 Dolgoprudny, Russia
[4] Institute for Information Transmission Problems (Kharkevich Institute) of Russian Academy of Science, 127051 Moscow, Russia; vladimir.voloshinov@gmail.com
[*] Correspondence: kukushkin_ab@nrcki.ru or kukushkin.alexander@gmail.com; Tel.: +7-499-196-7334

Received: 20 July 2018; Accepted: 8 August 2018; Published: 13 August 2018

Abstract: The accuracy of approximate automodel solutions for the Green's function of the Biberman-Holstein equation for the Stark broadening of spectral lines is analyzed using the distributed computing. The high accuracy of automodel solutions in a wide range of parameters of the problem is shown.

Keywords: Stark broadening of spectral lines; superdiffusive transport; distributed computing; automodel (self-similar) solutions

---

1. Introduction

The Stark broadening of spectral lines is known to produce the long-tailed spectral line shapes of atom and ion radiation in plasmas (see, e.g., [1–8]). In the broad range of conditions in plasmas and gases, where the complete redistribution (CRD) in photon frequency within the resonance line shape is applicable, the radiative transfer is described by the Biberman-Holstein equation for the density of excited atoms [9,10] and characterized by the infinite mean-squared displacement of the initial perturbation [11] and, respectively, by the irreducibility of the integral equation, in space variables, to a differential one. This makes the respective radiative transfer a nonlocal (superdiffusive) one [12,13] (for the deviation from the CRD and the limits of its applicability see, e.g., [14–16]).

The latter makes the numerical simulation of radiative transfer in resonance spectral lines a formidable task. The simple models based on the domination of the long-free-path flights of the photons were suggested [17] and developed for the quasi-steady-state transport problem, now known as the escape probability methods (see [18–20]).

For the time-dependent superdiffusive transport, recently, a wide class of non-steady-state superdiffusive transport on a uniform background with a power-law decay, at large distances, of the step-length probability distribution function (PDF) was shown [21] to possess an approximate automodel solution. The solution for the Green's function was constructed using the scaling laws for the propagation front (i.e., time dependence of the relevant-to-superdiffusion average displacement of the carrier) and asymptotic solutions far beyond and far ahead the propagation front. These scaling laws were shown to be determined essentially by the long free-path carriers (Lévy flights [22–26]). The validity of the suggested automodel solution was proved by its comparison with numerical solutions in the one-dimensional (1D) case of the transport equation with a simple long-tailed PDF with various power-law exponents and in the case of the Biberman-Holstein equation of the 3D resonance radiative transfer for various (Doppler, Lorentz, Voigt, and Holtsmark) spectral line shapes. The analysis of the limits of applicability of the automodel solution in the above-mentioned cases was continued in [27]. The full-scale numerical analysis of the limits of applicability was done in [28], using the massive computations of the exact solution of the transport equation with a simple long-tailed PDF with various power-law exponents. The comparison of these results with automodel solutions has shown high accuracy of automodel solutions in a wide range of space-time variables and enabled us to identify the limits of applicability of the automodel solutions.

It is important to note that the success of automodel solutions [21] for a wide class of non-stationary superdiffusive transport was achieved due to identification of the scaling laws in the case of radiative transfer in the Biberman-Holstein model in [11–13,17,21,29–33].

The Stark broadening of spectral lines is, itself, a highly complicated problem (see, e.g., [34] for the efforts, based on the quantum kinetic theory approach to the line broadening, to eliminate the remaining discrepancy of theory and experiment). The allowance for radiative transfer effects under conditions of the Stark broadening of spectral lines substantially complicates the diagnostics of the medium's parameters (see, e.g., the case of steady-state transport in inhomogeneous plasmas [35]). In light of these issues, the automodel solutions of the time-dependent Green's function of the radiative transfer equation in homogeneous media, considered in the present paper, may serve as reliable benchmarks for radiative transfer models in various applications.

Here we extend the approach [28] to the case of the Green's function of the 3D Biberman-Holstein equation for the Voigt spectral line shape, assuming the contribution of the impact Stark broadening to the Lorentz component of the Voigt line shape (Section 2.2.2). Additionally, the results for the case of Holtsmark line shape are presented (Section 2.2.3). The main equations are given in Section 2.1, and conclusions are made in Section 3.

2. Results

2.1. Main Equations

2.1.1. Biberman–Holstein Equation

The Biberman–Holstein equation [9,10] for radiative transfer in a uniform medium of two-level atoms/ions has been obtained from the system of equations for spatial density of excited atoms, $F(\mathbf{r}, t)$, and spectral intensity of resonance radiation. This system is reduced to a single equation for $F(\mathbf{r}, t)$, which appears to be an integral equation, non-reducible to a differential diffusion-type equation:

$$\frac{\partial F(\mathbf{r}, t)}{\partial t} = \frac{1}{\tau} \int_V G(|\mathbf{r} - \mathbf{r}_1|) F(\mathbf{r}_1, t) dV_1 - \left(\frac{1}{\tau} + \sigma\right) F(\mathbf{r}, t) + q(\mathbf{r}, t), \tag{1}$$

where τ is the lifetime of the excited atomic state with respect to spontaneous radiative decay; σ is the rate of the collisional quenching of excitation; q is the source of excited atoms, different from populating by the absorption of resonant photons (e.g., collisional excitation). The kernel G is determined by the (normalized) emission spectral line shape ε_v and the absorption coefficient k_v. In homogeneous media, G depends on the distance between the points of emission and absorption of the photon:

$$G(r) = -\frac{1}{4\pi r^2} \frac{dT(r)}{dr}, \quad T(r) = \int_0^\infty \varepsilon_v \exp(-k_v r) dv. \tag{2}$$

The analytical solution of Equation (1) with a point instant source, $q(\mathbf{r}, t) = \delta(\mathbf{r})\delta(t)$, i.e., the Green's function, was obtained in [11] with the help of the Fourier transform:

$$f(r, t) = -\frac{e^{-t(\frac{1}{\tau}+\sigma)}}{(2\pi)^2 r} \frac{\partial}{\partial r} \left\{ \int_{-\infty}^{\infty} e^{-ipr} \left[\exp\left\{ \frac{t}{\tau} J(p) \right\} - 1 \right] dp + 2\pi\delta(r) \right\}, \tag{3}$$

where:

$$J(p) = \frac{1}{p} \int_0^{\infty} \varepsilon_{\nu} k_{\nu} \arctg\frac{p}{k_{\nu}} d\nu. \tag{4}$$

The homogeneity of the medium implies that τ and σ are constants. The latter makes the role of quenching easily accounted for by the time exponent $\exp(-\sigma t)$, so in what follows we omit the account of this process, i.e., in fact, for convenience we take $\sigma = 0$. Hereafter we use the dimensionless time, assuming the normalization by τ.

Equation (3) for $r \neq 0$ may be transformed to take the following form:

$$f(r, t) = \frac{1}{(2\pi)^2 r} \int_{-\infty}^{\infty} p \sin(pr)[\exp\{t(J(p) - 1)\} - \exp\{-t\}]dp. \tag{5}$$

Let us consider two cases where the Stark broadening plays a dominant role in the resonance radiative transfer in the Biberman–Holstein model, namely, the case of the Voigt spectral line shape, assuming the contribution of the impact Stark broadening to the Lorentz component of the Voigt spectral line shape (Section 2.1.2), and the case of the static Stark broadening in the Holtsmark model (Section 2.1.3).

2.1.2. Exact Solution for Voigt Spectral Line Shape

The line shape ε_V is taken in the form [36]:

$$\varepsilon_V(a) = C'(a)\frac{2\sqrt{\ln 2}}{\Delta\nu_D} W(a, \omega(\nu)), \tag{6}$$

where:

$$W(a, \omega) = \int_{-\infty}^{+\infty} \frac{e^{-y^2} dy}{a^2 + (\omega - y)^2}, \tag{7}$$

$$\omega = \frac{2\sqrt{\ln 2}(\nu_0 - \nu)}{\Delta\nu_D}, \tag{8}$$

$$a = \frac{\Delta\nu_L}{\Delta\nu_D}\sqrt{\ln 2}. \tag{9}$$

Here $\Delta\nu_D$ is the full width at half maximum (FWHM) of the Doppler line shape, and $\Delta\nu_L$ is the FWHM of the Lorentz line shape. The coefficient $C'(a)$ is determined by the normalization condition, $\int_0^{\infty} \varepsilon_V d\nu = 1$. For the Stark broadening of the spectral line shape one may use the well-known formulae for the impact broadening, e.g., in the monographs [1,2], and learn the progress in the recent surveys [37,38].

The respective absorption coefficient $k_V(a)$ has the form:

$$k_V(a) = k_0 \frac{W(a, \omega(\nu))}{W(a, 0)},$$
$$k_0 = n_0 \pi \lambda^2 \frac{W(a, 0)}{\tau} \frac{g_n}{g_0}, \tag{10}$$

where n_0 is the density of absorbing atoms, λ, the wavelength of a photon, g_i, statistical weight of the i-th level. Using Equation (3.466.1) in [39], $W(a, 0)$ may be expressed in the form:

$$W(a, 0) = \int_{-\infty}^{+\infty} \frac{e^{-y^2} dy}{a^2 + y^2} = [1 - \text{erf}(a)] \frac{\pi}{a} e^{a^2}, \tag{11}$$

where $\text{erf}(x)$ is the error function:

$$\text{erf}(x) = \frac{2}{\sqrt{\pi}} \int_0^x e^{-t^2} dt. \tag{12}$$

Let us turn in Equation (5) to dimensionless variables $\rho = k_0 r$ and $P = p/k_0$, $v'' = 2\sqrt{\ln 2}(v - v_0)/\Delta v_D$ and use Equations (6) and (10) for ε_v and k_v. This gives:

$$G(\rho; a) = -k_0^3 \frac{1}{4\pi\rho^2} \frac{dT(\rho; a)}{d\rho}, \tag{13}$$

$$T(\rho; a) \equiv C_I(a) \int_{-\infty}^{\infty} W(a, -v'') \exp\left(-\frac{W(a, -v'')}{W(a, 0)}\rho\right) dv', \tag{14}$$

$$f(\rho, t; a) = k_0^3 \frac{1}{(2\pi)^2 \rho} \int_{-\infty}^{\infty} P \sin(P\rho)[\exp\{t(J(P; a) - 1)\} - \exp\{-t\}] dP, \quad \rho \neq 0, \tag{15}$$

$$J(P; a) = \frac{1}{P} \frac{C_I(a)}{W(a, 0)} \int_{-\infty}^{\infty} [W(a, -v'')]^2 \text{arctg} \frac{PW(a, 0)}{W(a, -v'')} dv''. \tag{16}$$

2.1.3. Exact Solution for Holtsmark Spectral Line Shape

For Holtsmark spectral line shape the functions ε_v and k_v, which enter the function $J(p)$, for linear Stark effect may be expressed in the form (cf. [3]):

$$\varepsilon_v = \frac{1}{\Delta v_H} \mathcal{H}\left[\frac{v - v_0}{\Delta v_H}\right], \quad \Delta v_H = C_2 F_0, \quad F_0 = 2.603 e N^{2/3}, \tag{17}$$

$$k_v = k_0 \mathcal{H}\left[\frac{v - v_0}{\Delta v_H}\right], \tag{18}$$

where N is the density of the perturbing particles, and $\mathcal{H}(\beta)$ is the Holtsmark function:

$$\mathcal{H}(\beta) = \frac{2}{\pi}\beta \int_0^{\infty} x \sin(\beta x) e^{-x^{3/2}} dx.$$

Turning in Equation (5) to dimensionless variables $\rho = k_0 r$ and $P = p/k_0$, $v' = (v - v_0)/\Delta v_H$ and using Equations (17) and (18) for ε_v and k_v, we obtain:

$$f(\rho, t) = -k_0^3 \frac{1}{(2\pi)^2 \rho} \frac{\partial}{\partial\rho} \int_{-\infty}^{\infty} \cos(P\rho)[\exp\{t(J(P) - 1)\} - \exp\{-t\}] dP, \quad \rho \neq 0, \tag{19}$$

$$J(P) = \frac{1}{P} \int_{-\infty}^{\infty} [\mathcal{H}(v')]^2 \text{arctg} \frac{P}{\mathcal{H}(v')} dv'. \tag{20}$$

2.2. Approximate Automodel Solution and Verification for Accuracy

2.2.1. General Equations

The approximate automodel solution [21] has the form:

$$f_{auto}(r,t) = tG\left(rg\left(\frac{r_{fr}(t)}{r}\right)\right),$$

(21)

where G is the kernel of the Biberman-Holstein equation, g is an unknown function with the following asymptotics:

$$g(s) = 1, \ s \ll 1,$$

(22)

$$g(s) \propto s, \ s \gg 1.$$

(23)

For the propagation front, $\rho_{fr}(t)$, we used in [21,28] the function which appears to be close to the time dependence of the mean displacement:

$$(t+1)T(\rho_{fr}(t)) = 1, \quad \rho = |x|.$$

(24)

Alternatively, one may use another function which appears to work better in the case of spectral line shape which is a convolution of two essentially different line shapes (e.g., for Voigt line shape):

$$f_{exact}(0,t) = t \, G(\rho_{fr}(t)),$$

(25)

where:

$$f_{exact}(0,t) = k_0^3 \frac{1}{(2\pi)^2} \int_{-\infty}^{\infty} P^2 [\exp\{t(J(P) - 1)\} - \exp\{-t\}] dP.$$

(26)

The relation between g and the exact solution of Equation (1), f_{exact}, is described by the following equation:

$$Q_G(\rho,t) \equiv \frac{1}{\rho} G^{-1}\left(\frac{f_{exact}(\rho,t)}{t}\right),$$

(27)

where G^{-1} is the function reciprocal to the G function, $\rho \equiv k_0 |r - r_0|$, k_0 is the absorption coefficient for photons at the frequency, corresponding to the line shape center:

$$Q_G(\rho,t(\rho,s)) \equiv Q_{G1}(s,\rho) = g(s),$$

(28)

$$Q_G(\rho(t,s),t) \equiv Q_{G2}(s,t) = g(s),$$

(29)

where the functions $t(\rho,s)$ and $\rho(t,s)$ are determined by the relation:

$$s = \frac{\rho_{fr}(t)}{\rho}.$$

(30)

To prove the automodel solution one has to show weak dependence (independence) of Q_{G1} and Q_{G2} functions on, respectively, the space coordinate and time. The results of the validation of the automodel solution and the reconstruction of function g from comparison of function (21) with computations of the Green's function for the Voigt and Holtsmark line shapes are given in what follows.

2.2.2. Automodel Solution for the Voigt Spectral Line Shape

The convolution of two line shapes with essentially different wings, an exponential one, for the Doppler case, and a power-law one, for the Lorentz case, makes the superdiffusive radiative transfer very sensitive to the contribution of the power-law wings of spectral line to the resulting long-tailed PDF. This is illustrated with the Figure 1 where the dependence of the exponent in the exact solution

(Equation (15)) is shown. It is seen that even a small fraction of the Lorentz line shape in the Voigt line shape produces strong effects at large distances (for small values of p, respectively).

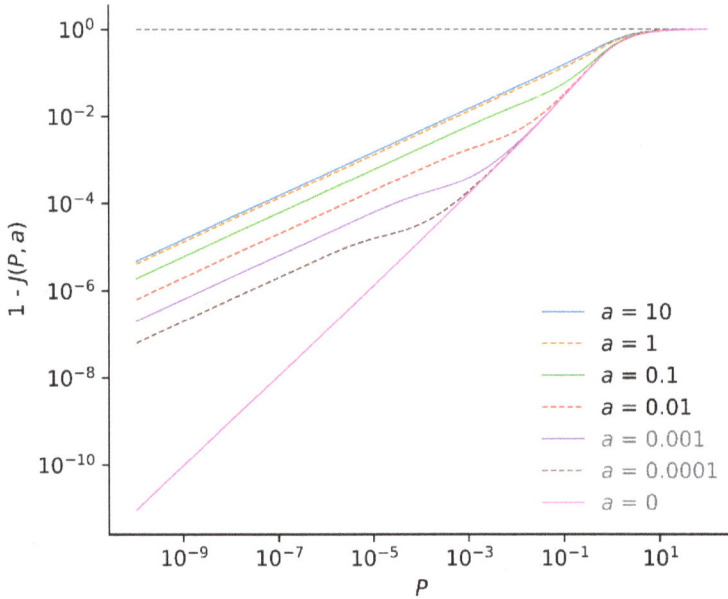

Figure 1. Dependence of the expression $(1 - J(P, a))$, where J is given by Equation (16) on the variable P for various values of a which is the Lorentz-to-Doppler width ratio in the Voigt spectral line shape.

We illustrate the results of accuracy analysis with three figures. First, the functions $Q_{G2}(s,t)$ (Equation (29)) are shown for different values of t in the range from $t_{min} = 30$ to $t_{max} = 10^6$. Further, the normalized functions $Q_{G2}(s,t)/\{Q_{G2}\}_{av}(s)$, where subscript av denotes averaging over time from $t_{min} = 30$ to $t_{max} = 10^8$, and the relative errors of the automodel solution $f_{auto}(\rho,t)/f_{exact}(\rho,t)$ are shown for the same range of time.

The results of analysis for various values of the Voigt parameter a are shown in Figure 2.

For applications it is of interest to determine the limits of applicability of the automodel solutions with a certain accuracy. To this end, in Figure 3 we show the level lines of the relative deviation of the automodel solution from the exact one, for the results from Figure 2.

It is seen that for the Lorentz-dominated line shapes, i.e., moderate and large values of the Voigt parameter a, the accuracy of automodel solution is high in almost the entire space of variables $\{\rho, t\}$, quite similar to the case [21,28] of the superdiffusive transport for a 1D model PDF with the power-law wings. However, for a small a, the accuracy dramatically falls down in the essential part of the space of variables $\{\rho, t\}$. This failure stimulated searching for improving the accuracy by using the propagation front (Equation (25)). The respective results are shown in Figures 4 and 5.

Figures 3 and 5 show that the automodel solution with the propagation front $\rho = \rho_{fr}$ (Equation (25)) substantially improves the accuracy in the regions of the failure of the automodel solution with the propagation front (Equation (24)). This issue points to the possibility to further improve the accuracy by the variation of the definition of the propagation front and further verification of the accuracy of the respective automodel function via calculations of the exact solution in a limited part of the space of variables $\{\rho, t\}$ (not in the entire space, only in the region around the propagation front).

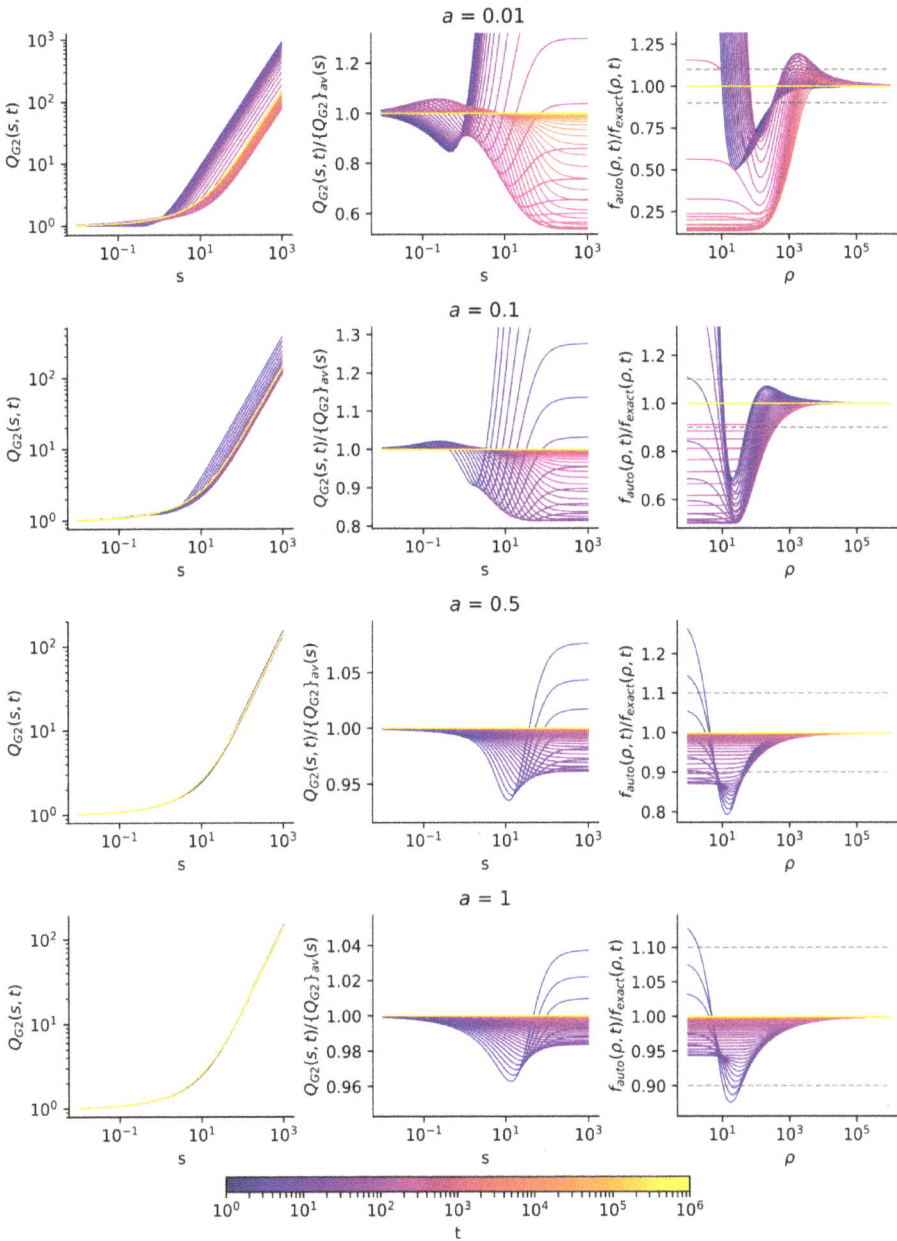

Figure 2. The result of accuracy analysis of automodel solution for various values of the Voigt parameter and the propagation front $\rho = \rho_{fr}$ taken in the form of Equation (24), for different values of t in the range from $t_{min} = 30$ to $t_{max} = 10^6$: (**left**) Functions $Q_{G2}(s,t)$ (29); (**center**) normalized functions $Q_{G2}(s,t)/\{Q_{G2}\}_{av}(s)$, where subscript av denotes averaging over time from $t_{min} = 30$ to $t_{max} = 10^8$; and (**right**) relative errors of the automodel solution $f_{auto}(\rho,t)/f_{exact}(\rho,t)$.

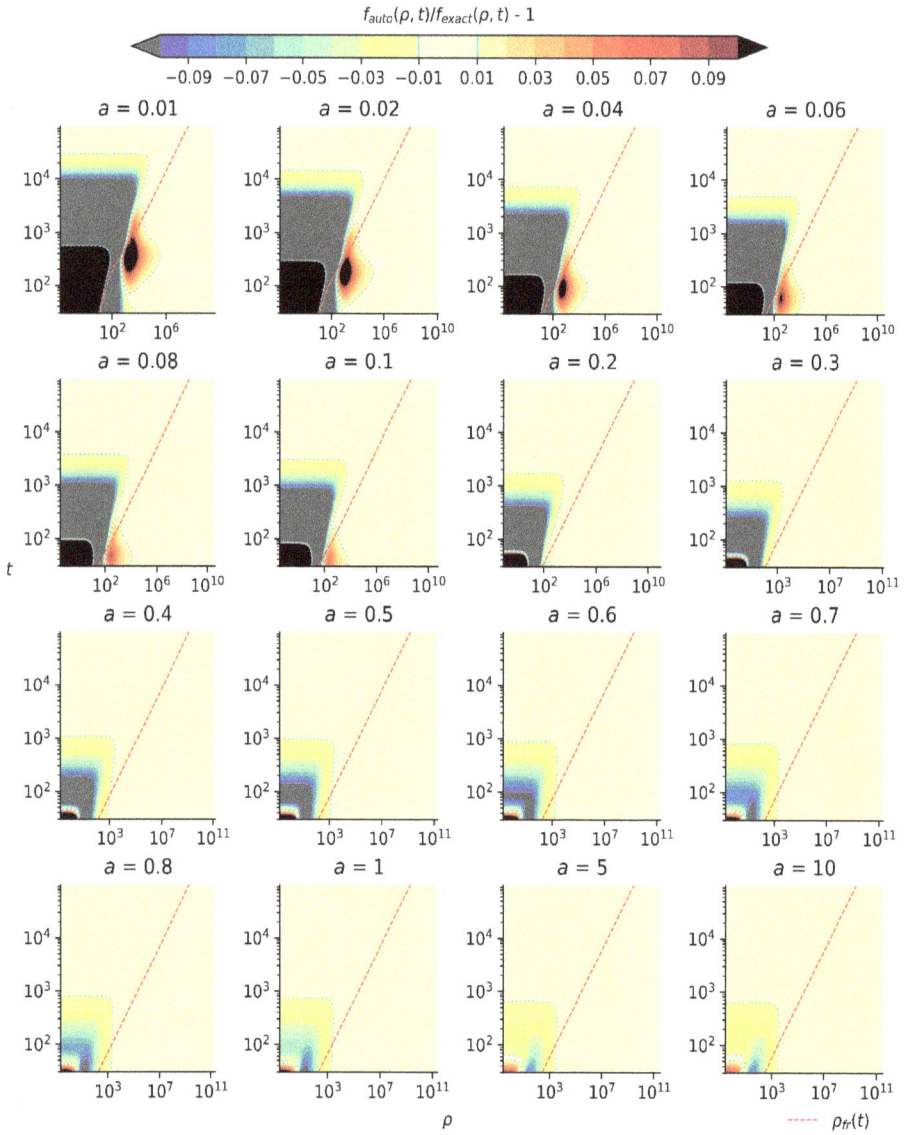

Figure 3. The level lines of the relative deviation of the automodel solution from the exact one, for the results from Figure 2.

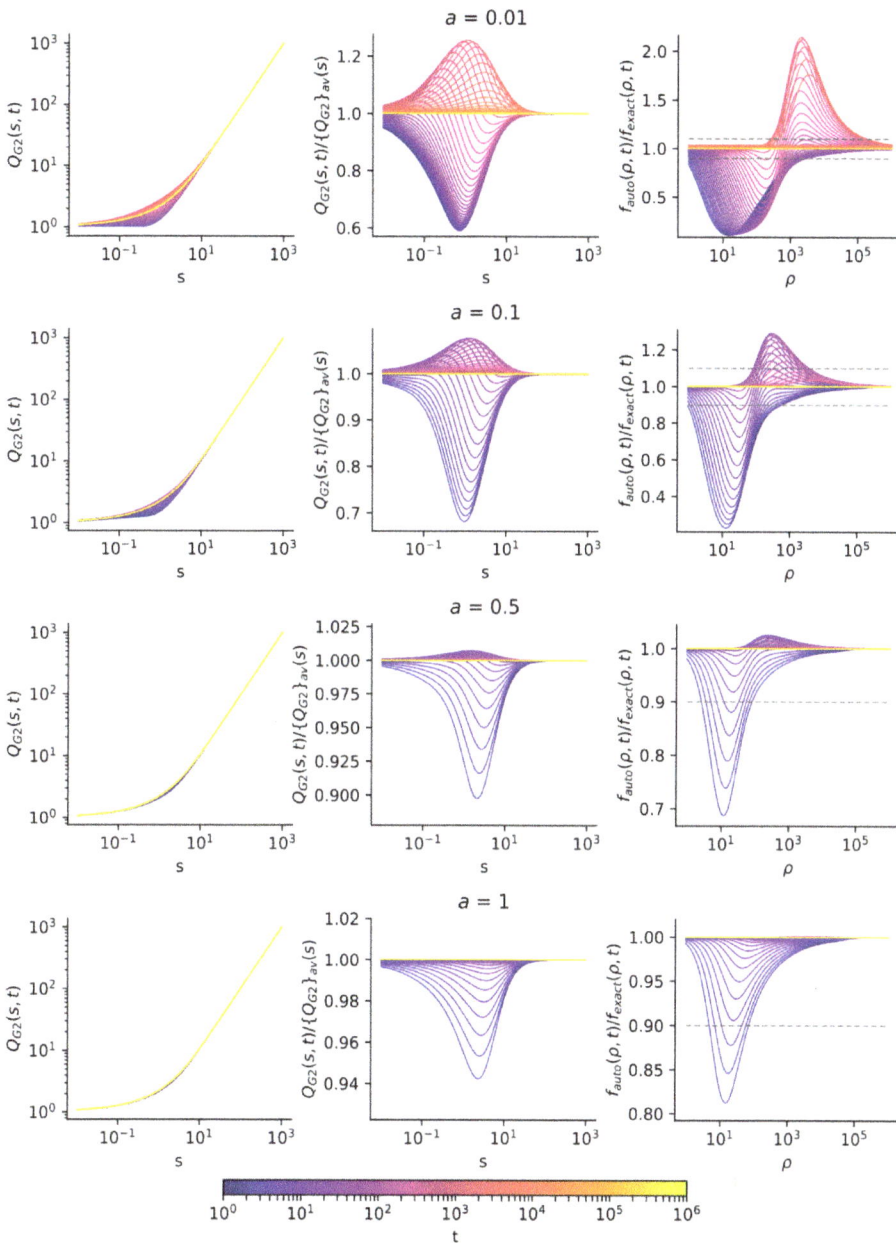

Figure 4. The same as in Figure 2, but for the propagation front $\rho = \rho_{\mathrm{fr}}$ taken in the form of Equation (25).

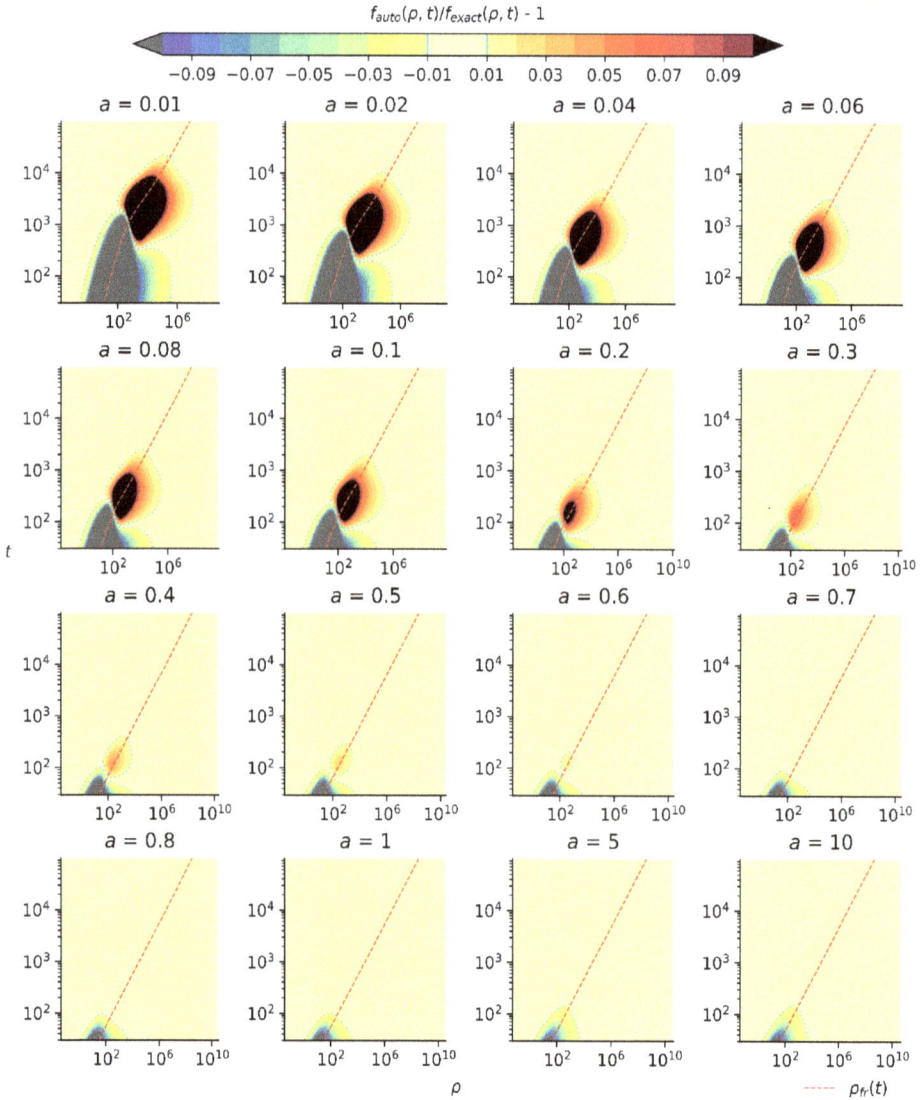

Figure 5. The same as in Figure 3, but for the propagation front $\rho = \rho_{\mathrm{fr}}$ taken in the form of Equation (25).

2.2.3. Automodel Solution for the Holtsmark Spectral Line Shape

Here we append the results [21,27] of analyzing the automodel solution for the Holtsmark line shape with the results for the accuracy of automodel solution. Figure 6 shows the behavior of the automodel function Q_{G2} (29) and the relative deviation of the automodel solution from the exact one in the range from $t_{\min} = 40$ to $t_{\max} = 10^3$, with the time step equal to 20, for $40 \leq t \leq 500$, and 50, for $500 < t \leq 1000$.

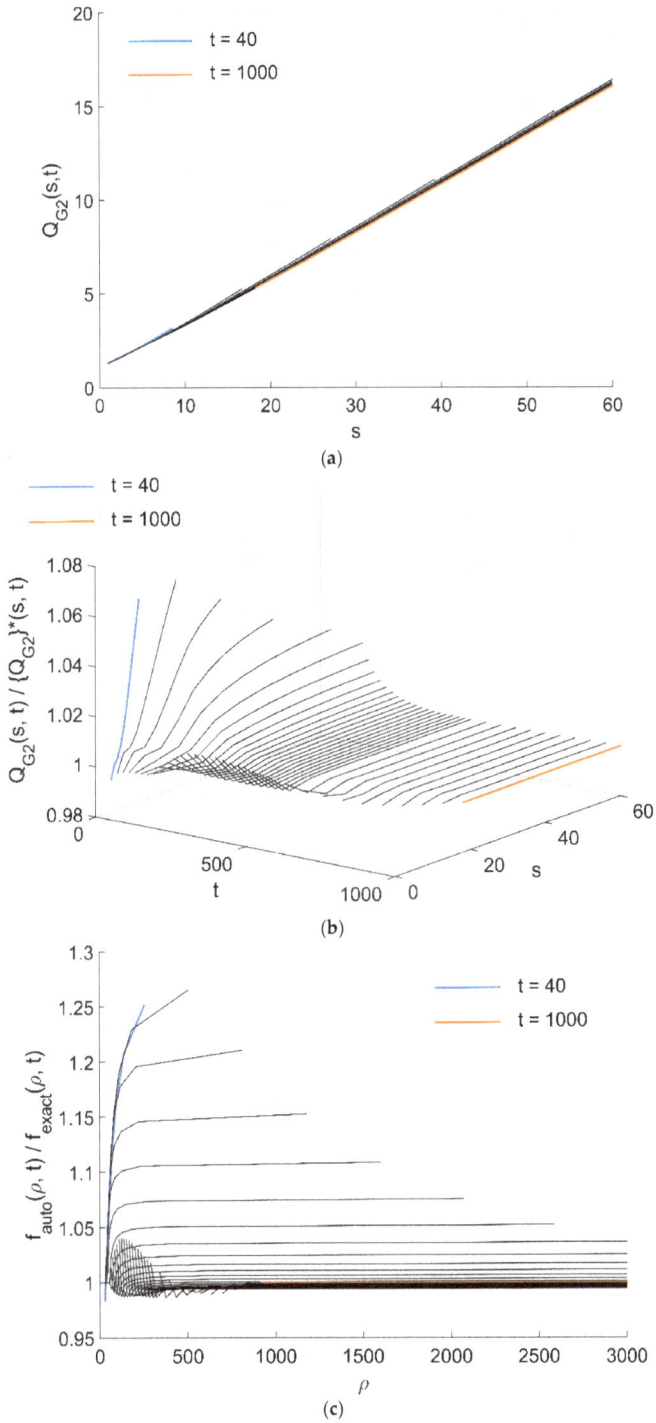

(a)

(b)

(c)

Figure 6. *Cont.*

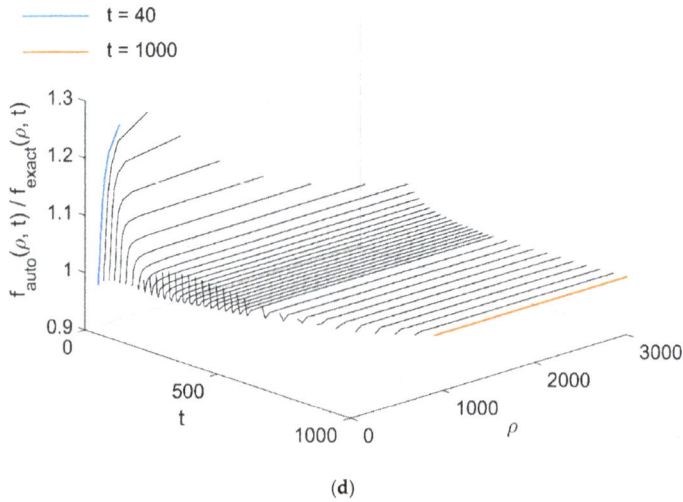

(d)

Figure 6. The result of accuracy analysis of automodel solution for the propagation front $\rho = \rho_{fr}$ taken in the form of Equation (24), for different values of t in the range from $t_{min} = 40$ to $t_{max} = 10^3$ with the time step equal to 20, if $40 \leq t \leq 500$. and 50, if $500 < t \leq 1000$: (**a**) functions $Q_{G2}(s,t)$ (29); (**b**) normalized functions $Q_{G2}(s,t)/\{Q_{G2}\}^*(s,t)$, where $\{Q_{G2}\}^*(s,t)$ is equal to $Q_{G2}(s,t = 160)$, for small s, and $Q_{G2}(s,t = 10^3)$, for large s; (**c**) relative error of the automodel solution $f_{auto}(\rho,t)/f_{exact}(\rho,t)$; and (**d**) the same in the 3D view.

The results show that, similarly to the 1D transport for a model step-length PDF with power-law wings [21,28], the accuracy of automodel solution is reasonably good for the propagation front (Equation (24)).

3. Discussion

The history of the research on the superdiffusive transport suggests that it is not possible to guess the "hidden" self-similarity from the general form of the transport equation, including the Biberman-Holstein equation, or from its analytical solutions in certain cases like the analytic solution [11] of the Biberman-Holstein equation. The approximate automodel (self-similar) solution [21] has been found thanks to:

- a hint from physics (namely, analysis of the kinetics of elementary excitation carriers); and
- interpolation of asymptotic solutions and solving an inverse problem which requires numerical simulations to verify the accuracy of the automodel solution and determine the limits of its applicability.

Obtaining automodel (self-similar) solutions in the entire space of independent variables requires mass numerical simulations, however, their total volume is significantly reduced due to the self-similarity of the solution.

The Stark broadening of spectral lines, including the contribution of the impact of Stark broadening to the Lorentz component of the Voigt line shape and the Holsmark broadening, produce the step-length probability distribution function (PDF) which makes the transport superdiffusive. The accuracy of approximate automodel solutions for the Green's function of the Biberman-Holstein equation for the Stark broadening of spectral lines is analyzed using the distributed computing. The high accuracy of automodel solutions in a wide range of parameters of the problem is shown. Massive computing experiments are conducted to verify automodel solutions. The Everest distributed computing platform and the cluster at the NRC 'Kurchatov Institute' are used. The results, obtained with distributed computing, verified the high accuracy of automodel solutions in a wide range of space-time variables and enabled us to identify the limits of applicability of automodel solutions.

The sensitivity of the automodel solution to the definition of the front propagation scaling is shown for the line shape with two quite different line broadening mechanisms, and the improvement of the accuracy of the automodel solution is achieved via generalization of the definition of the propagation front.

The present progress in describing the radiative transfer for infinite velocity of carriers, Equation (1), may be extended to the case of a finite velocity [40,41] (which includes the case of the resonance radiative transfer in astrophysics), where the exact solution and the asymptotics were also obtained [42].

4. Materials and Methods

Distributed computations have been done on the cluster at NRC "Kurchatov Institute" (http://ckp.nrcki.ru/) by means of the Everest (http://everest.distcomp.org/), a computing platform for publication, execution, and composition of applications running across distributed computing resources [43]. A generic Everest application, the so-called Parameter Sweep, (http://everest.distcomp.org/docs/ps) [44], was used to run a number of independent tasks at the cluster via a special Everest agent installed on the cluster. The calculation of exact and automodel solutions for a single value of parameter a in the Voight line shape (6) was an independent task. The values of a were taken in the range from 0.01 to 10, evenly spaced on a log scale, that yields 28 tasks. Since the two different expressions for the propagation front were considered (see Equations (24) and (25)) the total number of independent tasks was 56.

In order to obtain functions $g(s)$, the exact solutions $f_{exact}(\rho,t)$ were calculated. The values of t were evenly spaced on a log scale in the range from $t_{min} = 30$ to $t_{max} = 10^8$. The values of s were also evenly spaced on a log scale. Depending on the value of a, the lower limit for s was set in the range from 0.00003 (small a) to 0.005 (large a) while the upper limit was set to 1000 for all values of a. The respective values of ρ are defined by the t and s numerical meshes, using Equation (30).

All analytic expressions of Equations (6)–(30) have been implemented in the form of a Python program, using the NumPy and SciPy, well-known scientific libraries (www.scipy.org).

In addition, direct comparison of the automodel (Equation (21)) and exact (Equation (15)) solutions of the transport Equation (1), the range of variables in the two-dimensional space, {time, space coordinate}, where the automodel solution is accurate within certain error bars around the exact solution was identified.

Author Contributions: All authors contributed equally to this work.

Funding: This research was partially funded by Russian Foundation for Basic Research (RFBR), grant numbers 18-07-01269 A, 17-07-00950 A, 18-07-01175 A.

Acknowledgments: This work has been carried out using computing resources of the federal collective usage center Complex for Simulation and Data Processing for Mega-science Facilities at NRC "Kurchatov Institute" (ministry subvention under agreement RFMEFI62117X0016), http://ckp.nrcki.ru/. The authors are grateful to A.V. Demura, V.S. Lisitsa, and E.A. Oks for helpful discussions, and A.P. Afanasiev for the support of collaboration between the NRC "Kurchatov Institute" and the Center for Distributed Computing (http://distcomp.ru) of the Institute for Information Transmission Problems (Kharkevich Institute) of Russian Academy of Science.

Conflicts of Interest: The authors declare no conflict of interest. The funders had no role in the design of the study; in the collection, analyses, or interpretation of data; in the writing of the manuscript, and in the decision to publish the results.

References

1. Griem, H.R. *Principles of Plasma Spectroscopy*; Cambridge University Press: Cambridge, UK, 1997.
2. Sobel'man, I.I. *troduction to the Theory of Atomic Spectra*; Pergamon Press: Oxford, UK, 1972.

3. Kogan, V.I.; Lisitsa, V.S.; Sholin, G.V. Broadening of spectral lines in plasmas. In *Reviews of Plasma Physics*; Kadomtsev, B.B., Ed.; Consultants Bureau: New York, NY, USA, 1987; Volume 13, pp. 261–334.
4. Lisitsa, V.S. New results on the Stark and Zeeman effects in the hydrogen atom. *Sov. Phys. Usp.* **1987**, *30*, 927–951. [CrossRef]
5. Bureyeva, L.A.; Lisitsa, V.S. *A Perturbed Atom*; CRC Press: Boca Raton, FL, USA, 2000, ISBN 978-9058231383.
6. Oks, E. *Stark Broadening of Hydrogen and Hydrogenlike Spectral Lines in Plasmas: The Physical Insight*; Alpha Science International: Oxford, UK, 2006, ISBN 978-1842652527.
7. Oks, E. *Diagnostics Of Laboratory And Astrophysical Plasmas Using Spectral Lines Of One-, Two-, and Three-Electron Systems*; World Scientific: Hackensack, NJ, USA, 2017, ISBN 978-981-4699-07-5.
8. Demura, A.V. Beyond the Linear Stark Effect: A Retrospective. *Atoms* **2018**, *6*, 33. [CrossRef]
9. Biberman, L.M. On the diffusion theory of resonance radiation. *Sov. Phys. JETP* **1949**, *19*, 584–603.
10. Holstein, T. Imprisonment of Resonance Radiation in Gases. *Phys. Rev.* **1947**, *72*, 1212–1233. [CrossRef]
11. Veklenko, B.A. Green's Function for the Resonance Radiation Diffusion Equation. *Sov. Phys. JETP* **1959**, *9*, 138–142.
12. Biberman, L.M.; Vorob'ev, V.S.; Yakubov, I.T. *Kinetics of Nonequilibrium Low Temperature Plasmas*; Consultants Bureau: New York, NY, USA, 1987, ISBN 978-1-4684-1667-1.
13. Abramov, V.A.; Kogan, V.I.; Lisitsa, V.S. Radiative transfer in plasmas. In *Reviews of Plasma Physics*; Leontovich, M.A., Kadomtsev, B.B., Eds.; Consultants Bureau: New York, NY, USA, 1987; Volume 12, p. 151.
14. Starostin, A.N. Resonance radiative transfer. In *Encyclopedia of Low Temperature Plasma. Introduction Volume*; Fortov, V.E., Ed.; Nauka/Interperiodika: Moscow, Russia, 2000; Volume 1, p. 471. (In Russian)
15. Sechin, A.Y.; Starostin, A.N.; Zemtsov, Y.K.; Chekhov, D.I.; Leonov, A.G. Resonance radiation transfer in dense dispersive media. *J. Quant. Specrrosc. Radiat. Transf.* **1997**, *58*, 887–903. [CrossRef]
16. Bulyshev, A.E.; Demura, A.V.; Lisitsa, V.S.; Starostin, A.N.; Suvorov, A.E.; Yakunin, I.I. Redistribution function for resonance radiation in a hot dense plasma. *Sov. Phys. JETP* **1995**, *81*, 113–121.
17. Biberman, L.M. Approximate method of describing the diffusion of resonance radiation. *Doklady Akademii nauk SSSR Series Physics* **1948**, *49*, 659. (In Russian)
18. Napartovich, A.P. On the τ_{eff} method in the radiative transfer theory. *High Temp.* **1971**, *9*, 23–26.
19. Kalkofen, W. *Methods in Radiative Transfer*; Kalkofen, W., Ed.; Cambridge University Press: Cambridge, UK, 1984, ISBN 0-521-25620-8.
20. Rybicki, G.B. Escape Probability Methods. In *Methods in Radiative Transfer*; Kalkofen, W., Ed.; Cambridge University Press: Cambridge, UK, 1984; Chapter 1, ISBN 0-521-25620-8.
21. Kukushkin, A.B.; Sdvizhenskii, P.A. Automodel solutions for Lévy flight-based transport on a uniform background. *J. Phys. A Math. Theor.* **2016**, *49*, 255002. [CrossRef]
22. Shlesinger, M.; Zaslavsky, G.M.; Frisch, U. (Eds.) *Lévy Flights and Related Topics in Physics*; Springer-Verlag: Berlin, Germany, 1995, ISBN 978-3-662-14048-2.
23. Mandelbrot, B.B. *The Fractal Geometry of Nature*; W. H. Freeman: New York, NY, USA, 1982, ISBN 0-7167-1186-1189.
24. Eliazar, I.I.; Shlesinger, M.F. Fractional motions. *Phys. Rep.* **2013**, *527*, 101–129. [CrossRef]
25. Dubkov, A.A.; Spagnolo, B.; Uchaikin, V.V. Lévy flight superdiffusion: An introduction. *Int. J. Bifurc. Chaos* **2008**, *18*, 2649–2672. [CrossRef]
26. Klafter, J.; Sokolov, I.M. Anomalous diffusion spreads its wings. *Phys. World* **2005**, *18*, 29–32. [CrossRef]
27. Kukushkin, A.B.; Sdvizhenskii, P.A. Accuracy analysis of automodel solutions for Lévy flight-based transport: From resonance radiative transfer to a simple general model. *J. Phys. Conf. Series* **2017**, *941*, 012050. [CrossRef]
28. Kukushkin, A.B.; Neverov, V.S.; Sdvizhenskii, P.A.; Voloshinov, V.V. Numerical Analysis of Automodel Solutions for Superdiffusive Transport. *Int. J. Open Inf. Technol.* **2018**, *6*, 38–42.
29. Kogan, V.I. A Survey of Phenomena in Ionized Gases (Invited Papers). In Proceedings of the ICPIG'67, Vienna, Austria, 27 August–2 September 1968; p. 583. (In Russian)
30. Kogan, V.I. *Encyclopedia of Low Temperature Plasma. Introduction Volume*; Fortov, V.E., Ed.; Nauka/Interperiodika: Moscow, Russia, 2000; Volume 1, p. 481. (In Russian)
31. Abramov, Y.Y.; Napartovich, A.P. Transfer of resonance line radiation from a point source in the half-space. *Astrofizika* **1969**, *5*, 187–202. (In Russian)

32. Kukushkin, A.B.; Sdvizhenskii, P.A. Scaling Laws for Non-Stationary Biberman-Holstein Radiative Transfer. In Proceedings of the 2014 41st EPS Conference on Plasma Physics, Berlin, Germany, 23–27 June 2014.

33. Kukushkin, A.B.; Sdvizhenskii, P.A.; Voloshinov, V.V.; Tarasov, A.S. Scaling laws of Biberman-Holstein equation Green function and implications for superdiffusion transport algorithms. *Int. Rev. Atom. Mol. Phys.* **2015**, *6*, 31–41.

34. Iglesias, C.A. Electron broadening of isolated lines with stationary non-equilibrium level populations. *High Energy Density Phys.* **2005**, *1*, 42–51. [CrossRef]

35. Olson, G.L.; Comly, J.C.; La Gattuta, J.K.; Kilcrease, D.P. Stark broadened profiles with self-consistent radiation transfer and atomic kinetics in plasmas produced by high intensity lasers. *J. Quant. Spectrosc. Radiat. Transf.* **1994**, *51*, 255–261. [CrossRef]

36. Frish, S.E. *Optical Spectra of Atoms*; Fizmatgiz: Moscow-Leningrad, Russia, 1963. (In Russian)

37. Sahal-Bréchot, S.; Dimitrijevi'c, M.S.; Ben Nessib, N. Widths and shifts of isolated lines of neutral and ionized atoms perturbed by collisions with electrons and ions: An outline of the semiclassical perturbation (SCP) method and of the approximations used for the calculations. *Atoms* **2014**, *2*, 225–252. [CrossRef]

38. Stamm, R.; Hannachi, I.; Meireni, M.; Godbert-Mouret, L.; Koubiti, M.; Marandet, Y.; Rosato, J.; Dimitrijević, M.S.; Simić, Z. Stark Broadening from Impact Theory to Simulations. *Atoms* **2017**, *5*, 32. [CrossRef]

39. Gradshtein, I.S.; Ryzhik, I.M. *Tables of Integrals, Sums, Series and Products*; Fizmatgiz: Moscow, Russia, 1963; 1100p. (In Russian)

40. Zaburdaev, V.; Denisov, S.; Klafter, J. Lévy walks. *Rev. Mod. Phys.* **2015**, *87*, 483. [CrossRef]

41. Zaburdaev, V.Y.; Chukbar, K.V. Enhanced superdiffusion and finite velocity of Lévy flights. *J. Exp. Theor. Phys.* **2002**, *94*, 252–259. [CrossRef]

42. Kulichenko, A.A.; Kukushkin, A.B. Superdiffusive Transport of Biberman-Holstein Type for a Finite Velocity of Carriers: General Solution and the Problem of Automodel Solutions. *Int. Rev. Atom. Mol. Phys.* **2017**, *8*, 5–14.

43. Sukhoroslov, O.; Volkov, S.; Afanasiev, A. A Web-Based Platform for Publication and Distributed Execution of Computing Applications. In Proceedings of the 14th International Symposium on Parallel and Distributed Computing (ISPDC), Limassol, Cyprus, 29 June–2 July 2015; pp. 175–184.

44. Volkov, S.; Sukhoroslov, O. A Generic Web Service for Running Parameter Sweep Experiments in Distributed Computing Environment. *Procedia Comput. Sci.* **2015**, *66*, 477–486. [CrossRef]

atoms

MDPI

Article

Correcting the Input Data for Calculating the Asymmetry of Hydrogenic Spectral Lines in Plasmas

Paul Sanders * and Eugene Oks *

Physics Department, 206 Allison Lab, Auburn University, Auburn, AL 36849, USA
* Correspondence: phs0007@auburn.edu (P.S.); goks@physics.auburn.edu (E.O.)

Received: 17 February 2018; Accepted: 2 March 2018; Published: 6 March 2018

Abstract: We provide corrections to the data in Sholin's tables from his paper in *Optics and Spectroscopy* **26** (1969) 27. Since his data was used numerous times by various authors to calculate the asymmetry of hydrogenic spectral lines in plasmas, our corrections should motivate revisions of the previous calculations of the asymmetry and its comparison with the experimental asymmetry, and thus should have a practical importance.

Keywords: asymmetry of spectral lines; corrections to input data; hydrogenic spectral lines in plasmas

The input data presented in Sholin's tables from paper [1] was used numerous times by various authors to calculate the asymmetry of hydrogenic spectral lines in plasmas. (For the latest advances in the theory of the asymmetry we refer to papers [2,3] and references therein). However, we found that there are incorrect entries tabulated in paper [1] for the the Ly-γ, Ly-ϵ, and H-α lines, in both the intensity corrections and the quadrupole frequency corrections.

The dipole and quadrupole frequency corrections are given in paper [1] as

$$\Delta_k^{dipole} = nq - n'q',$$ (1)

and

$$\Delta_k^{quadrup} = \frac{1}{3}[n^4 - n^2 - 6n^2q^2 - n'^4 + n'^2 + 6n'^2q'^2],$$ (2)

where n and n' are the principal quantum numbers of the upper and lower energy levels, respectively; $q = n_1 - n_2$ and $q' = n_1' - n_2'$ are the combinations of the corresponding parabolic quantum numbers.

Frequency Corrections

For Ly-gamma ($n = 4$), Equation (2) becomes:

$$\Delta_k^{quadrup}(q) = 80 - 32q^2$$ (3)

It yields $\Delta_k^{quadrup}(0) = 80$, $\Delta_k^{quadrup}(\pm 1) = 48$, $\Delta_k^{quadrup}(\pm 2) = -48$, $\Delta_k^{quadrup}(\pm 3) = -208$. The comparison shows that in Sholin's table there are typographic errors in $\Delta_k^{quadrup}(0)$ entered as 60 (instead of 80) and in $\Delta_k^{quadrup}(\pm 3)$ entered as -206 (instead of -208).

For Ly-epsilon ($n = 4$), Equation (2) becomes:

$$\Delta_k^{quadrup}(q) = 420 - 72q^2$$ (4)

It yields $\Delta_k^{quadrup}(0) = 420$, $\Delta_k^{quadrup}(\pm 1) = 348$, $\Delta_k^{quadrup}(\pm 2) = 132$, $\Delta_k^{quadrup}(\pm 3) = -228$, $\Delta_k^{quadrup}(\pm 4) = -732$, $\Delta_k^{quadrup}(\pm 3) = -1380$. The comparison shows that in Sholin's table there are typographic errors in $\Delta_k^{quadrup}(\pm 2)$ entered as 108 (instead of 132).

Intensity Corrections

The intensity corrections are calculated from the corresponding corrections to the wave functions. The latter are given, e.g., in the Appendix of paper [4].

For H-alpha ($n = 3$ to $n = 2$ transition), the comparison shows that in Sholin's table there are typographic errors in $\epsilon_k^{(1)}$ corresponding to $\Delta_k^{dipole} = 2$ entered as -62 (instead of $-62/9$) and $\Delta_k^{dipole} = -2$ entered as 62 (instead of 62/9), as shown in detail below.

For $\Delta_k^{dipole} = 2$:

$$I_k = \; < 110''|z|010'' >^2$$
$$= (< 110|z|010 > -3\tfrac{a_\varrho}{R} < 200|z|010 > +3\tfrac{a_\varrho}{R} < 020|z|010$$
$$> -\tfrac{a_\varrho}{R} < 110|z|100 > +3\tfrac{a_\varrho^2}{R^2} < 200|z|100 > -3\tfrac{a_\varrho^2}{R^2} < 020|z|100 \qquad (5)$$
$$>)^2 \approx \; < 110|z|010 >^2 -6\tfrac{a_\varrho}{R} < 200|z|010 > +6\tfrac{a_\varrho}{R} < 020|z|010$$
$$> -2\tfrac{a_\varrho}{R} < 110|z|100 >= I_k{}^{(0)}\left(1 - \tfrac{a_\varrho}{R}\tfrac{62}{9}\right).$$

For $\Delta_k^{dipole} = -2$:

$$I_k = \; < 110''|z|100'' >^2$$
$$= (< 110|z|100 > -3\tfrac{a_\varrho}{R} < 200|z|100 > +3\tfrac{a_\varrho}{R} < 020|z|100$$
$$> +\tfrac{a_\varrho}{R} < 110|z|010 > -3\tfrac{a_\varrho^2}{R^2} < 200|z|010 > +3\tfrac{a_\varrho^2}{R^2} < 020|z|010 \qquad (6)$$
$$>)^2 \approx \; < 110|z|100 >^2 -6\tfrac{a_\varrho}{R} < 200|z|100 > +6\tfrac{a_\varrho}{R} < 020|z|100$$
$$> +2\tfrac{a_\varrho}{R} < 110|z|010 > = I_k{}^{(0)}\left(1 + \tfrac{a_\varrho}{R}\tfrac{62}{9}\right).$$

We note in passing that the robust perturbation theory, as developed by Oks and Uzer [5], allows for analytically calculating corrections to the eigenfunctions due to the quadrupole interaction in a much simpler way than in Sholin paper [1]. Details are presented in Appendix.

For completeness, we list below also previously known (for a long time) corrections to the tabulated entries from paper [1] for the H-beta line.

For the Stark components corresponding to the radiative transitions between the parabolic states 210 and 010 or between 120 and 100, the unperturbed intensity should be 81, instead of 16.

For the Stark component corresponding to the radiative transition between the parabolic states 210 and 001, the intensity correction $\epsilon_k^{(1)}$ should be -20 (instead of -16).

For the Stark component corresponding to the radiative transition between the parabolic states 120 and 001, the intensity correction $\epsilon_k^{(1)}$ should be 20 (instead of 16).

There are also two corrections (known for a long time) to the following typographic errors from paper [1].

In Table 2 from [1] for the H-alpha line, in the header of the last column, the scaling factor should be 10^6 instead of 10^5.

In Equation (21) from [1], in its 2nd term in the right hand side, the coefficient should be (3/8) instead of (3/16). We note that after this correction, Equation (21) from [1] coincides with the corresponding term (proportional to $1/R^4$) in Equation (4.59) from book [5] after setting in the latter $Z_1 = 1$, $Z_2 = Z$. Equation (4.59) from book [5] was derived from the exact expression for the energy in elliptical coordinates for the two Coulomb center problem by expanding the latter in powers of $1/R$ up to (including) the term $\sim 1/R^6$. Therefore, Equation (4.59) from book [5] can be considered, in particular, as the benchmark for testing Equation (21) from [1]. Such a test also confirms that the 2nd term in the right hand side of Equation (21) from [1] correctly contains the first power of Z (while there were incorrect suggestions that this term should contain Z^2).

In summary, since Sholin's input data from paper [1] was used numerous times by various authors to calculate the asymmetry of hydrogenic spectral lines in plasmas, our corrections should motivate revisions of the previous calculations of the asymmetry and its comparison with the experimental asymmetry, and thus should have a practical importance.

Acknowledgments: The authors are grateful to A.V. Demura for providing a valuable information.

Author Contributions: Both authors contributed equally to this work.

Conflicts of Interest: The authors declare no conflicts of interest.

Appendix A. Application of the Robust Perturbation Theory [5] for Calculating Quadrupole Corrections to the Wave Functions

Here, we use the robust perturbation theory [6]. The gist of it is as follows. If for the perturbed quantum system there is an operator A that commutes with the Hamiltonian H and the parts of these operators A_0 and H_0, characterizing the unperturbed quantum system, also commute, then the perturbation theory can be constructed in terms of the perturbation $(A - A_0)$ to the operator A_0, rather than in terms of the perturbation $(H - H_0)$ to the operator H_0. For calculating corrections to the wave functions (which are common for both A_0 and H_0), the advantage is that the eigenvalues of the operator A_0 are typically nondegenerate (in distinction to the eigenvalues of the operator H_0). Therefore, for calculating the first order corrections to the wave functions, it is sufficient to use the first order of the nondegenerate perturbation theory with respect to the perturbation $(A - A_0)$ and it would not involve infinite summations. In distinction, for calculating the same corrections in terms of the perturbation $(H - H_0)$, one would have to proceed to the second order of the degenerate perturbation theory, involving infinite summations.

Below as the operator A we choose the projection A_z of the super-generalized Runge-Lenz vector, derived by Kryukov and Oks [7], on the axis connecting the nucleus of the hydrogenic atom/ion with the perturbing ion. The operator of the unperturbed projection $A_z^{(0)}$ has the well-known eigenvalues q/n—see. e.g., the textbook [8]. According to Equation (12), from [6], the first non-vanishing term of the expansion of the operator $(A_z - A_z^{(0)})$ in terms of the small parameter n^2/R (here and below we use atomic units) is $-L^2/R$. Then, the corrections to the wave functions are given by

$$-\frac{1}{R}\frac{\left(L^2\right)_{nqm}^{nq'm}}{A_{z,a}{}^{(0)} - A_{z,a'}{}^{(0)}} = -\frac{n}{R}\frac{\left(L^2\right)_{nqm}^{nq'm}}{(q - q')}, \tag{A1}$$

where the selection rules for non-zero matrix elements of the operator L^2 require $q - q' = \pm 2$.

The non-diagonal matrix elements of the operator L^2 in parabolic coordinates (as well as of the operators $L_\pm = L_x \pm iL_y$), have been calculated by Sholin, Demura, and Lisitsa in [9]:

$$< n_1 + 1, n_2 - 1, m\left|L^2\right|n_1 n_2 m> = -[n_2(n - n_2)(n_1 + 1)(n - n_1 - 1)]^{1/2},$$
$$< n_1 - 1, n_2 + 1, m\left|L^2\right|n_1 n_2 m> = -[n_1(n - n_1)(n_2 + 1)(n - n_2 - 1)]^{1/2}. \tag{A2}$$

We note that matrix elements of the operator L_x in parabolic coordinates have been later reproduced by Gavrilenko in paper [10]. We also note that the non-diagonal matrix elements of the operators $L_\pm$ can also be obtained using their proportionality (within the manifold of the fixed n) to the non-diagonal matrix elements of the operators $(x \pm iy)$:

$$< n, q \pm 2, m|L_\pm|nqm> = -(\pm 1)[2/(3n)](x \pm iy). \tag{A3}$$

(The underlying physical reason for the existence of relation (A3) is, according to Demura [11], the O4 symmetry of hydrogenic atoms/ions.[1]) Therefore, the non-diagonal matrix elements of the operator L^2 in parabolic coordinates can be obtained using their similar proportionality to the non-diagonal matrix elements of the operator $(x^2 + y^2)$. The latter matrix elements have been calculated by Clark [12].

Anyway, after substituting the non-diagonal matrix elements of the operator L^2 from Equation (A2) in Equation (A1), the latter equation yields the following result for the corrections to the wave functions (more rigorously, for the coefficients of the corresponding linear combinations of the unperturbed wave functions):

$$\begin{aligned} \frac{n[n_2(n-n_2)(n_1+1)(n-n_1-1)]^{1/2}}{2R}, \quad q - q' = 2, \\ -\frac{n[n_1(n-n_1)(n_2+1)(n-n_2-1)]^{1/2}}{2R}, \quad q' - q = 2. \end{aligned} \tag{A4}$$

This is the same result as in Sholin paper [1], but it is obtained in a simpler way: without the need to go to the second order of the perturbation theory.

References

1. Sholin, G.V. On the nature of the asymmetry of the spectral line profiles of hydrogen in a dense plasma. *Opt. Spectrosc.* **1969**, *26*, 275–282.
2. Djurovic, S.; Ćirišan, M.; Demura, A.V.; Demchenko, G.V.; Nikolić, D.; Gigosos, M.A.; González, M.A. Measurements of H$_\beta$ Stark central asymmetry and its analysis through standard theory and computer simulations. *Phys. Rev. E* **2009**, *79*, 046402. [CrossRef] [PubMed]
3. Demura, A.V.; Demchenko, G.V.; Nikolic, D. Multiparametric dependence of hydrogen Stark profiles asymmetry. *Europ. Phys. J. D* **2008**, *46*, 111–127. [CrossRef]
4. Bacon, M.E. The asymmetry of Ly-α and Ly-β. *J. Quant. Spectrosc. Radiat. Transf.* **1976**, *17*, 501–512. [CrossRef]
5. Komarov, I.V.; Ponomarev, L.I.; Slavjanov, S.Y. *Spheroidal and Coulomb Spheroidal Functions*; Nauka: Moscow, Russia, 1976. (In Russian)
6. Oks, E.; Uzer, T. A robust perturbation theory for degenerate states based on the exact constants of the motion. *Europhys. Lett.* **2000**, *49*, 554–557. [CrossRef]
7. Kryukov, N.; Oks, E. Supergeneralized Runge-Lenz vector in the problem of two Coulomb or Newton centers. *Phys. Rev. A* **2012**, *85*, 054503. [CrossRef]
8. Landau, L.D.; Lifshitz, E.M. *Quantum Mechanics*; Pergamon: Oxford, UK, 1965.
9. Sholin, G.V.; Demura, A.V.; Lisitsa, V.S. Electron impact broadening of Stark sublevels of a hydrogen atom in a plasma. 1972, Preprint IAE-2232. (In Russian)
10. Gavrilenko, V.P. Resonant modification of quasistatic profiles of spectral lines of hydrogen in a plasma under the influence of noncollinear harmonic electric fields. *Sov. Phys. JETP* **1991**, *92*, 624–630.
11. Demura, A.V. Private communication, 2018.
12. Clark, C.W. Case of broken symmetry in the quadratic Zeeman effect. *Phys. Rev. A* **1981**, *24*, 605. [CrossRef]

---

[1] Specifically, this is related to the following two facts within the manifold of the fixed n [11]. First, the mean value <r> of the radius vector of the bound electron is proportional to the unperturbed Runge-Lenz vector $\mathbf{A}^{(0)}$, as it is well-known. Second, the linear combinations $\mathbf{J}_{\pm} = (\mathbf{L} \pm \mathbf{A}^{(0)})/2$ obey the same commutation relations as the angular momentum.

atoms

Article

Plasma Expansion Dynamics in Hydrogen Gas

Ghaneshwar Gautam [1] and Christian G. Parigger [2,*]

[1] Fort Peck Community College, 605 Indian Avenue, Poplar, MT 59255, USA; ggautam@fpcc.edu
[2] Department of Physics and Astronomy, University of Tennessee/University of Tennessee Space Institute, 411 B.H. Goethert Parkway, Tullahoma, TN 37388, USA
* Correspondence: cparigge@tennessee.edu; Tel.: +1-931-841-5690

Received: 13 July 2018; Accepted: 10 August 2018; Published: 20 August 2018

Abstract: Micro-plasma is generated in ultra-high-pure hydrogen gas, which fills the inside of a cell at a pressure of $(1.08 \pm 0.033) \times 10^5$ Pa by using a Q-switched neodymium-doped yttrium-aluminum-garnet (Nd:YAG) laser device operated at a fundamental wavelength of 1064 nm and a pulse duration of 14 ns. The micro-plasma emission spectra of the hydrogen Balmer alpha line, H_α, are recorded with a Czerny–Turner type spectrometer and an intensified charge-coupled device. The spectra are calibrated for wavelength and corrected for detector sensitivity. During the first few tens of nanoseconds after the initiation of optical breakdown, the significant Stark-broadened and Stark-shifted H_α lines mark the well-above hypersonic outward expansion. The vertical diameters of the spectrally resolved plasma images are measured for the determination of expansion speeds, which were found to decrease from 100 to 10 km/s for time delays of 10 to 35 ns. For time delays of 0.5 μs to 1 μs, the expansion speed of the plasma decreases to the speed of sound of 1.3 km/s in the near ambient temperature and pressure of the hydrogen gas.

Keywords: laser–plasma interactions; plasma dynamics and flow; hypersonic flows; Emission Spectra

1. Introduction

In laser-induced plasma, the ambient gas can foster or diminish the plasma expansion. The pressure and type of gases also influences the post-breakdown phenomena. For example, at low pressures, the losses and uniformity of the plasma energy distribution increases [1]. Plasma size, propagation speed and emission properties are also related to the ambient gas into which the plasma expands. The physical cause of the high-speed expansion is the large pressure difference between the plasma and its surrounding environment.

This work investigates the expansion dynamics of plasma in hydrogen gas at a pressure that is slightly above 10^5 Pa. The study of the plasma expansion is applicable for astrophysical, engineering and scientific research as well as other various applications [2–5]. The astrophysical interests include the interpretation of white dwarf photosphere absorption spectra by simulating the conditions of these selected stars in laboratory settings [2–4] and by exploring laser-induced plasma [4] with time-resolved laser spectroscopy. The motivation behind this engineering and scientific research [5] extends to the measurement of phenomena that occur at hypersonic speeds to re-entry speeds that are in the Mach number range of 5–25 or above.

During the plasma expansion, the spatial and temporal variations of density, temperature and pressure are observed. Investigations of these spatial and temporal profiles allow us to infer the expansion speeds. Plasma at elevated temperatures and pressures expands at a high velocity and causes a shock wave that propagates into the outward direction. As the incident laser beam is focused at a point above the breakdown threshold irradiance, the breakdown occurs at a location before the pulse reaches its focal point [6].

The interaction between the laser beam and the material is a complex process, which depends on many characteristics, such as laser parameters or target materials. Various factors affect this interaction, including the properties of the pulse, such as the pulse width, spatial and temporal fluctuation of the pulse, as well as the peak irradiance variations. For example, the effects of the pulse width on nascent laser-induced bubbles for underwater laser-induced breakdown spectroscopy (LIBS) show that a long pulse causes well-defined spectra with clear lines. In turn, a short pulse usually causes considerably asymmetric or deformed spectra. However, this effect is more significant for a solid target material than for a gaseous target material [7]. Moreover, the images of laser-induced breakdown plasma in air [8] nicely illustrate expansion dynamics.

The experimental results in air, which were obtained using a Nd:YAG laser at the wavelength of 532 nm and the pulse width of 6.5 ns [8], show two distinct regions with higher intensity towards the laser propagation direction for time delays of 25 ns to 10 μs. Schlieren images of laser-induced plasma generated in air at the standard ambient temperature and pressure also show plasma jet propagation towards the laser for time delays of 1–20 μs [9]. However, the jet propagation direction depends on the type of gas, the gas pressure, the ratio of energy absorbed in the plasma and the threshold irradiance for the occurrence of optical breakdown [9].

2. Results

This section elaborates on the determination of the expansion speed of the micro-plasma, which was generated in the ultra-high-pure hydrogen gas inside a cell. Experimental results show high plasma expansion speeds that decreased from 100 to 10 km/s at early time delays of 10 ns to 35 ns.

Figure 1 displays the recorded H_α plasma spectra at early time delays of 10 ns to 35 ns, which were measured at 5-ns time intervals. The 2-dimensional spectra of slit-height versus wavelength are significantly Stark-broadened and Stark-shifted at early time delays. The measured intensity increases for successive time delays. However, the area and line-width of the spectral profiles decrease continuously, which implies decreasing electron density. Based on the recorded H_α plasma spectra at early time delays as displayed in Figure 1, plasma expansion speeds were determined. The diameter of the plasma in the lateral or slit-height direction is measured as a function of time and hence, the plasma expansion-speed can be determined. For example, the red arrows on the spectra images at 10 ns and 15 ns in Figure 1 indicate the spatial plasma ranges used for the determination of the expansion speeds.

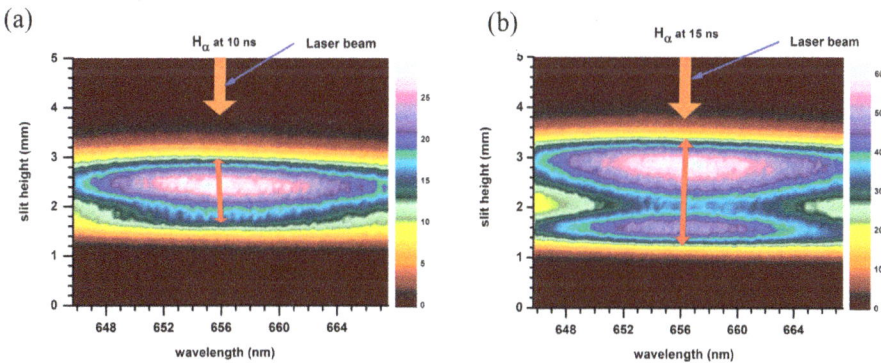

Figure 1. *Cont.*

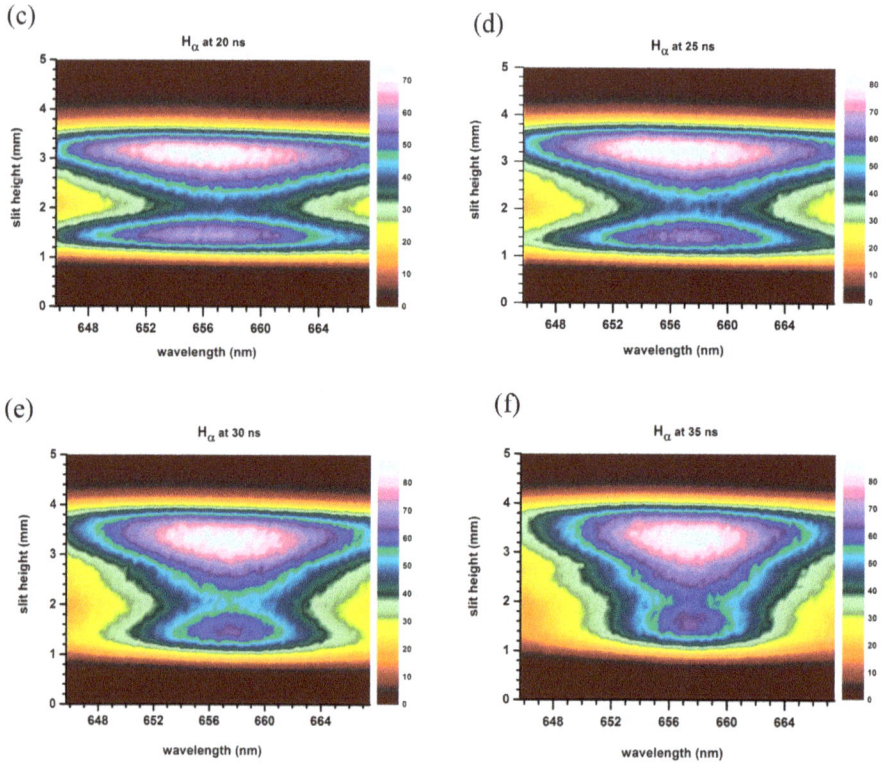

Figure 1. Hydrogen alpha plasma spectra images for selected time delays: (**a**) 10 ns; (**b**) 15 ns; (**c**) 20 ns; (**d**) 25 ns; (**e**) 30 ns; and (**f**) 35 ns.

The recorded images of the laser-induced plasma with 14-ns pulses depict a higher intensity towards the laser at early time delays of 10 ns to 35 ns. However, the opposite behavior is observed at later time delays, e.g., for time delays of 400 ns. Figure 2 illustrates the experimental records for time delays of 400 and 900 ns.

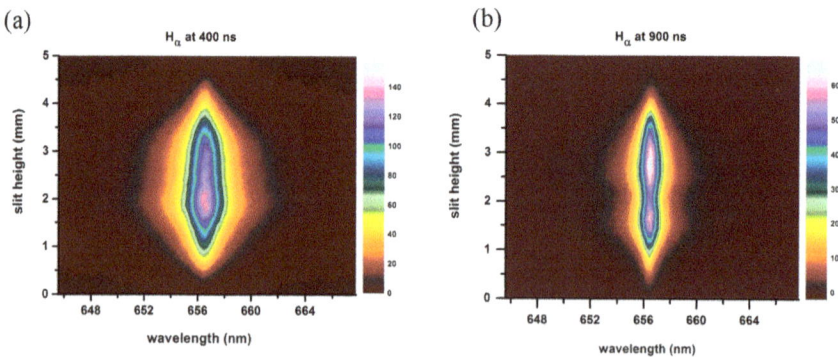

Figure 2. Hydrogen alpha plasma spectra images at time delays of (**a**) 400 ns and (**b**) 900 ns.

From the images in Figure 1, the diameter of the plasma in the lateral or slit-height direction is measured as a function of time and hence, plasma expansion speeds can be determined. Table 1 displays the measured diameters of the plasma and the corresponding speeds at various time delays.

Table 1. Plasma expansion speed at various time delays using lateral direction expansion from the contour plots exhibited in Figures 1 and 2.

Time (ns)	Diameter (mm)	Distance in 5 ns (mm)	Speed (km/s)
10	1.32	--	--
15	2.03	0.71	142
20	2.42	0.39	78
25	2.71	0.29	58
30	2.86	0.15	30
35	2.92	0.06	12
400	0.96	--	--
900	1.81	0.85 *	1.70

* Distance is for 500 ns.

The predicted plasma expansion speeds decrease from 100 to 10 km/s for time delays of 15 to 35 ns. The determined expansion speeds are well above the hypersonic speed (Mach number 5) or above the re-entry speeds (Mach number 25) at these time delays. Figure 3 illustrates the tabulated results.

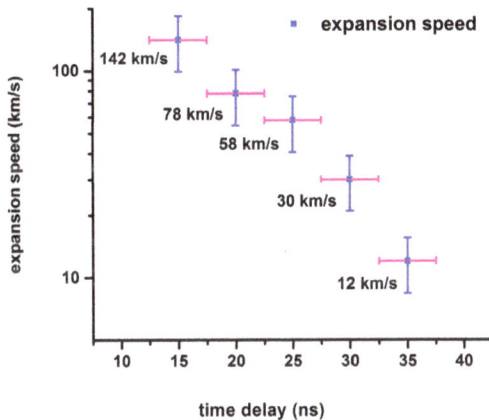

Figure 3. Plasma expansion speeds in log scale (see Table 1). The indicated time-delay error bars are due to the gate width of 5 ns.

The plasma expansion typically decreases to hypersonic and sonic speeds for larger time delays as indicated in Figure 3. The measured expansion speeds are consistent with the results from previous experimental studies [10,11]. However, computer simulations show a shock-wave expansion speed of 60 km/s at a time delay of 20 ns in air at atmospheric pressure [12], thereby indicating that the determined speeds are consistent with other experimental results. It is important to note that the speed of sound in hydrogen gas is 1.3 km/s, which is 3.5-fold higher than in the ambient pressure and temperature of the air.

For longer time delays of about 1 μs, the images in Figure 2 are utilized for the determination of the plasma expansion speed. The determined speed is 1.7 ± 0.5 km/s at 900 ns. The estimated error bars for these speeds are ± 30%. These results agree well with those of the recent hydrogen experiments [13].

3. Materials and Methods

During the experiment, a cell is filled with ultra-high-pure hydrogen gas at a pressure of $(1.08 \pm 0.033) \times 10^5$ Pa. To study time-resolved and space-resolved emission spectroscopy, a Q-switched Nd:YAG laser device (Quanta Ray DCR-2A (10), Spectra-Physics, Santa Clara, CA, USA [14]) is used at its fundamental wavelength of 1064 nm with 10-Hz repetition rate and full-width-half-maximum pulses of 14 ns. The measured energy per pulse is 120 mJ. The laser beam was passed through a dichroic beam splitter to remove the residual 532-nm component. A silicon photodiode detector was used to record a portion of the laser radiation that is reflected off of the beam splitter at the exit of the laser source. The photodiode is connected to an oscilloscope to monitor the optical pulse. Three mirrors (NB1-K13; Thorlabs, Newton, NJ, USA [4]) are used to align the beam to be parallel to the spectrometer slit. Figure 4 illustrates the experimental schematic.

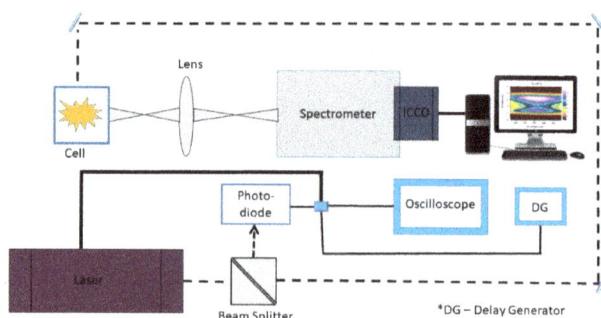

Figure 4. Experimental schematic for time-resolved laser spectroscopy.

A holographic grating of 1200 grooves/mm is selected to disperse the radiation from the plasma. For the recording of temporally and spatially resolved plasma emission spectra images along the slit height, the following instrumentation is employed: Czerny–Turner type spectrometer (0.64-m HR640; Jobin-Yvon, Longjumeau, FR [4]) and intensified charge-coupled device (ICCD) (Andor technology model iStar DH334T-25U-03, South Windsor, CT, USA [4]). The spectral resolution was 0.11 nm. The data were recorded with a 5-ns gate width and an average of 100 consecutive laser–plasma events were accumulated. For later time delays of 400 ns and 900 ns, 50 consecutive laser–plasma events were accumulated using a 20-ns gate width. The recorded spectra are wavelength calibrated and corrected for detector sensitivity. Matlab® (Mathworks, Nattick, MA, USA [4]) scripts are utilized for wavelength calibration and sensitivity correction, while Origin Software (OriginLab, Northampton, MA, USA [4]) is employed for the graphical display of the experimental data.

4. Conclusions

The laser-induced plasma expands at a well-above hypersonic speed depending upon the ambient conditions and time delays. The spectra recorded during the evolution of the plasma are significantly Stark-broadened and Stark-shifted. Therefore, a larger percentage error occurs for the predicted speeds for earlier time delays. For longer time delays, the plasma expansion speeds decrease considerably and thus, larger differences occur for the speed measurement due to the increased temporal interval. To improve the graphically inferred expansion speeds, Abel inversion methods can be applied so that radial information will be extracted from the recorded line-of-sight measurements [14]. The predicted expansion speeds may be useful for the National Aeronautics and Space Administration (NASA) hypersonic technology (HT) project. Details of the focal volume irradiance distribution [15] for the study of laser-induced optical breakdown may augment future experimental studies.

Atoms **2018**, *6*, 46

Author Contributions: C.G.P. and G.G. designed and performed the experiments. G.G. analyzed the result and wrote the paper with suggestions from C.G.P.

Funding: This work is in part supported by the Center for Laser Applications, a State of Tennessee supported Center of Excellence, at The University of Tennessee Space Institute.

Conflicts of Interest: The authors declare no conflict of interest.

References

1. Anabitarte, F.; Cobo, A.; Lopez-Higuera, J.M. Laser-induced breakdown spectroscopy: Fundamentals, applications, and challenges. *Int. Sch. Res. Notices Spectrosc.* **2012**, *2012*, 285240. [CrossRef]
2. Falcon, R.E.; Rochau, G.A.; Bailey, J.E.; Ellis, J.L.; Carlson, A.L.; Gomez, T.A.; Montgomery, M.H.; Winget, D.E.; Chen, E.Y.; Gomez, M.R.; et al. An experimental platform for creating white dwarf photospheres in the laboratory. *High Energy Density Phys.* **2013**, *9*, 82–90. [CrossRef]
3. Falcon, R.E.; Rochau, G.A.; Bailey, J.E.; Gomez, T.A.; Montgomery, M.H.; Winget, D.E.; Nagayama, T. Laboratory measurements of white dwarf photospheric spectral lines: H_β. *Astrophys. J.* **2015**, *806*, 214. [CrossRef]
4. Parigger, C.G.; Drake, K.A.; Helstern, C.M.; Gautam, G. Laboratory hydrogen-beta emission spectroscopy for analysis of astrophysical white dwarf spectra. *Atoms* **2018**, *6*, 36. [CrossRef]
5. Engeln, R.; Mazouffre, S.; Vankan, P.; Bakker, I.; Schram, D.C. Plasma expansion: fundamentals and applications. *Plasma Sources Sci. Technol.* **2002**, *11*, A100. [CrossRef]
6. Chen, Y.L.; Lewis, J.W.L.; Parigger, C.G. Spatial and temporal profiles of pulsed laser-induced air plasma emissions. *J. Quant. Spectrosc. Radiat. Transf.* **2000**, *67*, 91–103. [CrossRef]
7. Sakka, T.; Tamura, A.; Matsumoto, A.; Fukami, K.; Nishi, N.; Thornton, B. Effects of pulse width on nascent width on nascent laser-induced bubbles for underwater laser-induced breakdown spectroscopy. *Spectrochim. Acta Part B At. Spectrosc.* **2014**, *97*, 94–98. [CrossRef]
8. Glumac, N.; Elliott, G.; Boguszko, M. Temporal and spatial evolution of a laser spark in air. *Am. Inst. Aeronaut. Astronaut. J.* **2005**, *34*, 1984–1994. [CrossRef]
9. Brieschenk, S.; O'Byrne, S.; Kleine, H. Visualization of jet development in laser-induced plasmas. *Opt. Lett.* **2013**, *38*, 664–666. [CrossRef] [PubMed]
10. Parigger, C.G.; Plemmons, D.H.; Lewis, J.W.L. Spatially and temporally resolved electron number density measurements in a decaying laser-induced plasma using hydrogen-alpha line profiles. *Appl. Opt.* **1995**, *34*, 3325–3330. [CrossRef] [PubMed]
11. Parigger, C.G. Atomic and molecular emissions in laser-induced breakdown spectroscopy. *Spectrochim. Acta Part B At. Spectrosc.* **2013**, *79*, 4–16. [CrossRef]
12. Sobral, H.; Villagrán-Muniz, M.; Navarro-González, R.; Raga, A.C. Temporal evolution of the shock wave and hot core air in laser induced plasma. *Appl. Phys. Lett.* **2000**, *77*, 3158–3160. [CrossRef]
13. Parigger, C.G.; Surmick, D.M.; Gautam, G. Self-absorption characteristics of measured laser-induced plasma line shapes. *J. Phys. Conf. Ser.* **2017**, *810*, 012012. [CrossRef]
14. Parigger, C.G.; Gautam, G.; Surmick, D.M. Radial electron density measurements in laser-induced plasma from Abel inverted hydrogen Balmer beta line profiles. *Int. Rev. At. Mol. Phys.* **2015**, *6*, 43–55.
15. Parigger, C.G. Laser-induced breakdown in gases: Experiments and simulation. In *Laser-Induced Breakdown Spectroscopy*; Miziolek, A.W., Palleschi, V., Schechter, I., Eds.; Cambridge University Press: Cambridge, UK, 2006; ISBN 978-0521071000.

Communication

X-ray Spectroscopy Based Diagnostic of GigaGauss Magnetic Fields during Relativistic Laser-Plasma Interactions

Elisabeth Dalimier [1] and Eugene Oks [2],*

[1]   LULI—Sorbonne Université-Campus Pierre et Marie Curie, CNRS, Ecole Polytechnique,
       CEA: Université Paris-Saclay, CEDEX 05, F-75252 Paris, France; elisabeth.dalimier@upmc.fr
[2]   206 Allison Lab, Physics Department, Auburn University, Auburn, AL 36849, USA
*   Correspondence: goks@physics.auburn.edu

Received: 20 September 2018; Accepted: 1 November 2018; Published: 6 November 2018

Abstract: GigaGauss (GG), and even multi-GG magnetic fields are expected to be developed during relativistic laser-plasma interactions. Sub-GG magnetic fields were previously measured by a method using the self-generated harmonics of the laser frequency, and the fact that the magnetized plasma is birefringent and/or optically active depending on the propagation direction of the electromagnetic wave. In the present short communication, we outline an idea for a method of measuring GG magnetic fields based on the phenomenon of Langmuir-wave-caused dips (L-dips) in X-ray line profiles. The L-dips were observed in several experimental spectroscopic studies of relativistic laser-plasma interactions. Ultrastrong magnetic fields affect the separation of the L-dips from one another, so that this relative shift can be used to measure such fields.

Keywords: relativistic laser-plasma interactions; GigaGauss magnetic fields; X-ray spectral line profiles; Langmuir-wave-caused dips

---

GigaGauss (GG), and even multi-GG magnetic fields are expected to be developed during relativistic laser-plasma interactions. These fields should be localized at the surface of the relativistic critical density—see, e.g., review [1] and references therein. In particular, according to Equation (11) from paper [2], the maximum magnetic field B_{max} is related to the laser intensity I as follows:

$$B_{max} \ (G) = 10^{-1}[I(W/cm^2)]^{1/2}. \tag{1}$$

So, at the laser intensities I ~10^{21} W/cm^2 achieved in recent experiments (see paper [3]), the magnetic fields can be as high as B_{max} ~3 GG.

On the experimental side, in paper [4] magnetic fields B ~0.7 GG were measured by using the polarization measurements (the Cotton-Mouton effect of an induced ellipticity) of high-order VUV laser harmonics generated at the incident irradiation intensity $I = 9 \times 10^{19}$ W/cm^2. In an earlier experiment [5,6], magnetic fields up to B ~0.4 GG were measured at the incident irradiation intensity up to $I = 9 \times 10^{19}$ W/cm^2, by a method also using the self-generated harmonics of the laser frequency and the fact that the magnetized plasma is birefringent (the Cotton-Mouton effect) and/or optically active (the Faraday effect of the rotation of the polarization vector) depending on the propagation direction of the electromagnetic wave.

In the present short communication, we propose a method for measuring GG magnetic fields based on the phenomenon of Langmuir-wave-cased dips (L-dips) in X-ray line profiles. The L-dips were observed in several experimental spectroscopic studies of relativistic laser-plasma interactions—see, e.g., papers [3,7] and review [8].

According to the theory (presented, e.g., in books [9,10]), L-dips originate from a dynamic resonance between the Stark splitting

$$\omega_{stark}(F) = 3n\hbar F/(2Z_r m_e e) \tag{2}$$

of hydrogenic energy levels, caused by a quasistatic part of the electric field F in a plasma, and the frequency ω_L of the Langmuir wave, which practically coincides with the plasma electron frequency $\omega_{pe} = (4\pi e^2 N_e/m_e)^{1/2}$:

$$\omega_{stark}(F) = s\omega_{pe}\ (N_e),\ s = 1, 2, \ldots \tag{3}$$

Here n and Z_r are the principal quantum number and the nuclear charge of the radiating hydrogenic atom/ion (radiator), s is the number of quanta (Langmuir plasmons) involved in the resonance. Despite the applied electric field being quasimonochromatic, there occurs a nonlinear dynamic resonance of a multifrequency nature, as explained in detail in paper [11].

From the resonance condition (3), one determines the specific locations of L-dips in spectral line profiles, which depend on N_e, since ω_{pe} depends on N_e. Generally, there could be two sets of L-dips in the spectral line profile at distances $\Delta\omega_{dip}$ from the unperturbed frequency ω_0 of the spectral line. One set, located at

$$\Delta\omega_{dip}{}^{(\alpha)} = (q_\alpha - q_\beta n_\beta/n_\alpha)s\omega_{pe} \tag{4}$$

results from the resonance with the splitting of the upper sublevel α (of the principal quantum number n_α) involved in the radiative transition. Another set located at

$$\Delta\omega_{dip}{}^{(\beta)} = (q_\alpha n_\alpha/n_\beta - q_\beta)s\omega_{pe} \tag{5}$$

results from the resonance with the splitting of the lower sublevel β (of the principal quantum number n_β) involved in the radiative transition. Here $q = n_1 - n_2$ is the electric quantum number expressed via the parabolic quantum numbers n_1 and n_2: $q = 0, \pm 1, \pm 2, \ldots, \pm(n-1)$. The electric quantum numbers mark Stark components of hydrogenic spectral lines. It should be emphasized that for the Ly-lines, there is no second set of the L-dips at $\Delta\omega_{dip}{}^{(\beta)}$ because there is no linear Stark splitting of the state of $n = 1$. Below for brevity we omit the subscript "pe" and use ω instead of ω_{pe}.

In paper [12], for the specific case of the one-quantum resonance ($s = 1$) in hydrogen atoms ($Z_r = 1$), Gavrilenko generalized Equations (4) and (5) for the situation where there is also a magnetic field B in plasmas. His corresponding formulas are as follows:

$$\Delta\omega_{dip}{}^{(\alpha)} = \omega\{(n' + n'')_\alpha - [(n'+n'')_\beta/n_\alpha][(n_\alpha{}^2 - n_\beta{}^2)b_0{}^2 + n_\beta{}^2]^{1/2}\}, \tag{6}$$

$$\Delta\omega_{dip}{}^{(\beta)} = \omega\{[(n' + n'')_\alpha/n_\beta][\ n_\alpha{}^2 - (n_\alpha{}^2 - n_\beta{}^2)b_0{}^2]^{1/2} - (n' + n'')_\beta\}. \tag{7}$$

Here the quantum numbers n' and n'' correspond to the basis of the wave functions diagonalizing the Hamiltonian of a hydrogen atom in a non-collinear static electric (F) and magnetic (B) fields (see, e.g., paper [13]):

$$n', n'' = -j, -j+1, \ldots, j; j = (n-1)/2. \tag{8}$$

The quantity b_0 in Equations (6) and (7) is the scaled, dimensionless magnetic field

$$b_0 = \mu_0 B/(\hbar\omega), \tag{9}$$

where μ_0 is the Bohr magneton.

We further slightly generalize Gavrilenko's formulas by allowing for any number of quanta s involved in the resonance and for any nuclear charge Z_r of hydrogenic atoms/ions:

$$\Delta\omega_{dip}{}^{(\alpha)} = s\omega\{(n' + n'')_\alpha - [(n' + n'')_\beta/n_\alpha][(n_\alpha{}^2 - n_\beta{}^2)b^2 + n_\beta{}^2]^{1/2}\}, \tag{10}$$

$$\Delta\omega_{dip}{}^{(\beta)} = s\omega\{[(n' + n'')_\alpha/n_\beta][\ n_\alpha{}^2 - (n_\alpha{}^2 - n_\beta{}^2)b^2]^{1/2} - (n' + n'')_\beta\}, \tag{11}$$

where the scaled dimensionless magnetic field b now reads:

$$b = \mu_0 B/(s\hbar\omega) = (1/s)[B(GG)/0.201][\omega(s^{-1})/(1.77 \times 10^{15})]^{-1} \tag{12}$$

For example, for the one-quantum resonance ($s = 1$), for the frequency $\omega = 1.77 \times 10^{15}$ s^{-1}, which is the frequency of the laser used, e.g., in experiments [3,7], the quantity b reaches unity at $B = 0.201$ GG. We note that the nuclear charge Z_r does not enter Equations (10) and (11), but obviously does affect the unperturbed frequency of the spectral line.

The idea of a new method for measuring the magnetic fields is as follows. It is possible to select such a pair of the L-dip at $\Delta\omega_{dip}{}^{(\alpha)}$ and the L-dip at $\Delta\omega_{dip}{}^{(\beta)}$, both corresponding to the same combination of the sums $(n' + n'')_\alpha$ and $(n' + n'')_\beta$, such that the location of one of the two L-dips is unaffected by the magnetic field while the location of the other of the two L-dips is shifted by the magnetic field. Then from the relative separation of the two L-dips it is possible to determine the magnetic field.

Namely, we are talking about the following pairs of the L-dips. One pair corresponds to

$$(n' + n'')_\alpha = 0, (n' + n'')_\beta = -1, \tag{13}$$

while another pair corresponds to

$$(n' + n'')_\alpha = 1, (n' + n'')_\beta = 0. \tag{14}$$

The ratio

$$\Delta\omega_{dip}{}^{(\alpha)}/\Delta\omega_{dip}{}^{(\beta)} = (1/n_\alpha)[(n_\alpha{}^2 - n_\beta{}^2)b^2 + n_\beta{}^2]^{1/2} \tag{15}$$

in the first case and the ratio

$$\Delta\omega_{dip}{}^{(\beta)}/\Delta\omega_{dip}{}^{(\alpha)} = (1/n_\beta)[n_\alpha{}^2 - (n_\alpha{}^2 - n_\beta{}^2)b^2]^{1/2} \tag{16}$$

in the second case are simple functions of the magnetic field, as it is seen from the above formulas.

Figure 1 shows the ratio $\Delta\omega_{dip}{}^{(\alpha)}/\Delta\omega_{dip}{}^{(\beta)}$ in the pair of the L-dips corresponding to $(n' + n'')_\alpha = 0, (n' + n'')_\beta = -1$, versus the scaled dimensionless magnetic field b for the Balmer-alpha line (solid curve) and for the Balmer-beta line (dashed curve).

dip positions ratio

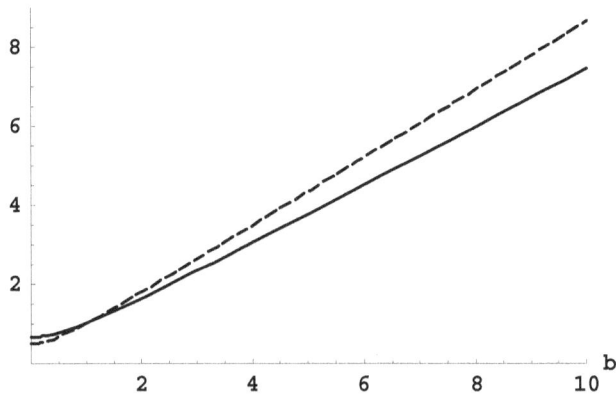

Figure 1. The ratio of positions $\Delta\omega_{dip}{}^{(\alpha)}/\Delta\omega_{dip}{}^{(\beta)}$ in the pair of the L-dips corresponding to $(n' + n'')_\alpha = 0, (n'+n'')_\beta = -1$, versus the scaled (dimensionless) magnetic field b (defined by Equation (12)) for the Balmer-alpha line (solid curve) and for the Balmer-beta line (dashed curve).

It is seen that in the range of b presented in Figure 1, the magnetic field significantly affects the relative positions of the L-dips, so that by measuring the latter it is possible to determine the magnetic field. For the laser frequency $\omega = 1.77 \times 10^{15}$ s^{-1}used, e.g., in experiments [3,7], the range of $b \sim (1\text{--}10)$ corresponds to the range of the magnetic field $B \sim (0.2\text{--}2)$ GG for the one-quantum resonance and to $B \sim (0.4\text{--}4)$ GG for the two-quantum resonance. For $b >> 10$, the possible L-dips at $\Delta\omega_{dip}^{(\alpha)}$ would be shifted too far into the wings of the spectral lines, so that most probably they could not be observed.

For completeness we note that if one would use the pair of the L-dips in the profiles of Stark components characterized by the quantum numbers from Equation (14), then according to Equation (16) the range of b would be limited to $b_{max} = n_\alpha^2/(n_\alpha^2 - n_\beta^2)$. This is because at $b_{max} = n_\alpha^2/(n_\alpha^2 - n_\beta^2)$, the possible L-dips at $\Delta\omega_{dip}^{(\beta))}$ would disappear.

Here is a practical example based on measuring the relative shift of the L-dips in the profiles of the Balmer lines of Cu XXIX. (We note that it is technologically simple to make and use thin Cu foils to irradiate them by a powerful laser). The wavelengths of the Balmer-alpha and Balmer-beta lines of Cu XXIX are 0.77 nm and 0.57 nm, respectively. This is practically the same range of the wavelength as it was employed, e.g., in experiments [3,7] while studying the L-dips in the profiles of the Ly-beta line of Si XIV and Al XIII. Therefore, the same kind of spectrometers can be used without any major additional tuning for experimental studies of possible L-dips in the profile of the Balmer-alpha and Balmer-beta lines of Cu XXIX, and thus for the experimental determination of GG (or sub-GG) magnetic fields.

In summary, ultrastrong magnetic fields affect the separation of the L-dips from one another, so that this relative shift can be used to measure sub-GG and GG magnetic fields. Earlier there was proposed another diagnostic of magnetic fields in plasmas based on the polarization measurements of X-ray spectral line profiles [14]. However, the method proposed in the present research note is easier to implement experimentally: it does not require performing the polarization measurements in the X-ray range, which would be relatively difficult to implement.

Author Contributions: Both authors contributed equally.

Funding: This work has been done within the LABEX Plas@par project. It received a state financial support by the Agence Nationale de la Recherche, as a part of the program "Investissements d'avenir" under the reference ANR-11-IDEX-0004-02.

Conflicts of Interest: The authors declare no conflict of interest.

References

1.   Belyaev, V.S.; Krainov, V.P.; Lisitsa, V.S.; Matafonov, A.P. Generation of fast charged particles and superstrong magnetic fields in the interaction of ultrashort high-intensity laser pulses with solid targets. *Physics-Uspekhi* **2008**, *51*, 793. [CrossRef]

2.   Belyaev, V.S.; Matafonov, A.P. Fast Charged Particles and Super-Strong Magnetic Fields Generated by Intense Laser Target Interaction. In *Femtosecond-Scale Optics*; Andreev, A., Ed.; InTech: Shanghai, China, 2011.

3.   Oks, E.; Dalimier, E.; Faenov, A.Y.; Angelo, P.; Pikuz, S.A.; Tubman, E.; Butler, N.M.H.; Dance, R.J.; Pikuz, T.A.; Skobelev, I.Y.; et al. Using X-ray spectroscopy of relativistic laser plasma interaction to reveal parametric decay instabilities: a modeling tool for astrophysics. *Opt. Express* **2017**, *25*, 1958–1972. [CrossRef] [PubMed]

4.   Wagner, U.; Tatarakis, M.; Gopal, A.; Beg, F.N.; Clark, E.L.; Dangor, A.E.; Evans, R.G.; Haines, M.G.; Mangles, S.P.D.; Norreys, P.A.; et al. Laboratory measurements of 0.7 GG magnetic fields generated during high-intensity laser interactions with dense plasmas. *Phys. Rev. E* **2004**, *70*, 026401. [CrossRef] [PubMed]

5.   Tatarakis, M.; Gopal, A.; Watts, I.; Beg, F.N.; Dangor, A.E.; Krushelnick, K. Measurements of ultrastrong magnetic fields during relativistic laser–plasma interactions. *Phys. Plasmas* **2002**, *9*, 2244. [CrossRef]

6.   Tatarakis, M.; Watts, I.; Beg, F.N.; Clark, E.L.; Dangor, A.E.; Gopal, A.; Haines, M.G.; Norreys, P.A.; Wagner, U.; Wei, M.-S.; et al. Measuring huge magnetic fields. *Nature* **2002**, *415*, 280. [CrossRef] [PubMed]

7.   Oks, E.; Dalimier, E.; Faenov, A.Y.; Angelo, P.; Pikuz, S.A.; Pikuz, T.A.; Skobelev, I.Y.; Ryazanzev, S.N.; Durey, P.; Doehl, L.; et al. In-depth study of intra-Stark spectroscopy in the X-ray range in relativistic laser–plasma interactions. *J. Phys. B At. Mol. Opt. Phys.* **2017**, *50*, 245006. [CrossRef]

8. Dalimier, E.; Pikuz, T.; Angelo, P. Mini-Review of Intra-Stark X-ray Spectroscopy of Relativistic Laser–Plasma Interactions. *Atoms* **2018**, *6*, 45. [CrossRef]
9. Oks, E. *Plasma Spectroscopy: The Influence of Microwave and Laser Fields*; Springer Series on Atoms and Plasmas; Springer: New York, NY, USA, 1995; Volume 9.
10. Oks, E. *Diagnostics of Laboratory and Astrophysical Plasmas Using Spectral Lines of One-, Two-, and Three-Electron Systems*; World Scientific: Hackensack, NJ, USA, 2017.
11. Gavrilenko, V.P.; Oks, E. New effect in Stark spectroscopy of atomic hydrogen: dynamic resonance. *Sov. Phys. JETP* **1981**, *53*, 1122.
12. Gavrilenko, V. Resonance effects in the spectroscopy of atomic hydrogen in a plasma with a quasimonochromatic electric field and located in a strong magnetic field. *Sov. Phys. JETP* **1988**, *67*, 915.
13. Demkov, Y.; Monozon, B.; Ostrovsky, V. Energy levels of a hydrogen atom in crossed electric and magnetic fields. *Sov. Phys. JETP* **1970**, *30*, 775–776.
14. Demura, A.V.; Oks, E. New method for polarization measurements of magnetic fields in dense plasmas. *IEEE Trans. Plasma Sci.* **1998**, *26*, 1251–1258. [CrossRef]

MDPI

St. Alban-Anlage 66

4052 Basel

Switzerland

Tel. +41 61 683 77 34

Fax +41 61 302 89 18

www.mdpi.com

Atoms Editorial Office

E-mail: atoms@mdpi.com

www.mdpi.com/journal/atoms

www.ingramcontent.com/pod-product-compliance
Lightning Source LLC
Chambersburg PA
CBHW041216220326
41597CB00033BA/5988

* 9 7 8 3 0 3 8 9 7 4 5 5 0 *